lonely planet KIDS

EXPLORAPEDIA

AMAZING EXPLORERS OF THE WORLD

AND THEIR JOURNEYS OF DISCOVERY

ACKNOWLEDGEMENTS

Author: Emma Marriott
Illustrator: Michelle Pereira
Publisher: Piers Pickard
Art Director: Andy Mansfield
Editorial Director: Joe Fullman
Commissioning Editor: Kate Baker
Consultant: Dr John Haywood
Print Production: Nigel Longuet

Published in October 2022 by Lonely Planet Global Ltd

CRN: 554153 • ISBN: 978 1 83869 519 4
www.lonelyplanet.com/kids • © Lonely Planet 2022

Printed in Singapore • 10 9 8 7 6 5 4 3 2 1

STAY IN TOUCH – LONELYPLANET.COM/CONTACT

Ireland: Digital Depot, Roe Lane (off Thomas St), Digital Hub, Dublin 8, D08 TCV4

Paper in this book is certified against the Forest Stewardship Council™ standards. FSC™ promotes environmentally responsible, socially beneficial and economically viable management of the world's forests.

EXPLORAPEDIA

AMAZING EXPLORERS OF THE WORLD

AND THEIR JOURNEYS OF DISCOVERY

EMMA MARRIOTT & MICHELLE PEREIRA

Contents

6 **INTRODUCTION**

First Journeys

8 **OCEAN EXPLORERS**

10 *Ancient Seafarers*

12 *Polynesian Navigators*

14 *Viking Explorers*

16 Leif Erikson (c.970–c.1020) and Gudrid Thorbjarnardóttir (980–1019)

18 Zheng He (1371–1433)

20 *The Age of Discovery*

22 *Mapping the World*

24 Christopher Columbus (1451–1506)

26 Vasco da Gama (c.1460–1524) and Ahmad Ibn Mājid (c.1432–c.1500)

28 Ferdinand Magellan (1480–1521)

30 Hasekura Rokuemon Tsunenaga (1571–1622)

32 Jeanne Baret (1740–1807)

34 Captain James Cook (1728–1779)

36 Bungaree (c.1775–1830) and Matthew Flinders (1774–1814)

38 Charles Darwin (1809–1882)

40 *Ocean Record-Breakers*

42 *The Ocean Deep*

44 Jacques Cousteau (1910–1997) and Jacques Piccard (1922–2008)

46 Marie Tharp (1920–2006) and Dr Cindy Lee Van Dover (1954–)

48 **OVERLAND EXPLORERS**

50 Marco Polo (1254–1324) and Rabban Bar Sauma (1220–1294)

52 Ibn Battuta (1304–1369)

54 *Conquistadors*

56 Estevanico (c.1500–1539)

58 Francisco de Orellana (1511–1546)

60 Meriwether Lewis (1774–1809), William Clark (1770–1838) and Sacagawea (c.1788–1812 or 1884)

62 James Beckwourth (c.1798–c.1866)

64 *The Railway Age*

66 Isabella Bird (1831–1904)

68 Robert O'Hara Burke (1821–1861) and William John Wills (1834–1861)

70 *Songlines*

72 *African Adventures*

74 Mary Kingsley (1862–1900) and Florence Baker (1841–1916)

76 Gertrude Bell (1868–1926)

78 *Rediscovering Old Civilisations*

80 The Villas-Bôas Brothers: Orlando (1914–2002), Cláudio (1916–1998), Leonardo (1918–1961)

82 Nellie Bly (1864–1922)

84 *Around the World by Any Means*

86 **EXPLORERS OVER ICE AND SNOW**

88 *The Search for the Northwest Passage*
90 Matthew Henson (1866–1955) and Robert Peary (1856–1920)
92 *The Race to the South Pole*
94 Robert Falcon Scott (1868–1912) and Roald Amundsen (1872–1928)
96 Ernest Shackleton (1874–1922)
98 Ranulph Fiennes (1944–) and Ann Bancroft (1955–)
100 *Mountain Climbers*
102 Edmund Hillary (1919–2008) and Tenzing Norgay (1914–1986)

104 **AIR AND SPACE EXPLORERS**

106 *Fearless Flyers*
108 Louis Blériot (1872–1936) and Harriet Quimby (1875–1912)
110 Charles Lindbergh (1902–1974) and Amelia Earhart (1897–1939)
112 *The Space Race*
114 Yuri Gagarin (1934–1968) and Valentina Tereshkova (1937–)
116 Neil Armstrong (1930–2012) and the Apollo Astronauts
118 *Space Pioneers*
120 *Future Exploration*

122 **AN INTERVIEW WITH A MODERN-DAY EXPLORER**

Emily Ford, the first woman and first person of colour to hike the Ice Age Trail

124 Glossary
126 Index

Humans are explorers, driven to see what lies beyond the horizon. Curiosity or competition can push us to explore, and sometimes our very survival depends on it. That urge to discover the unknown drove early humans thousands of years ago to venture out of Africa and to settle on new continents across the globe.

First Journeys

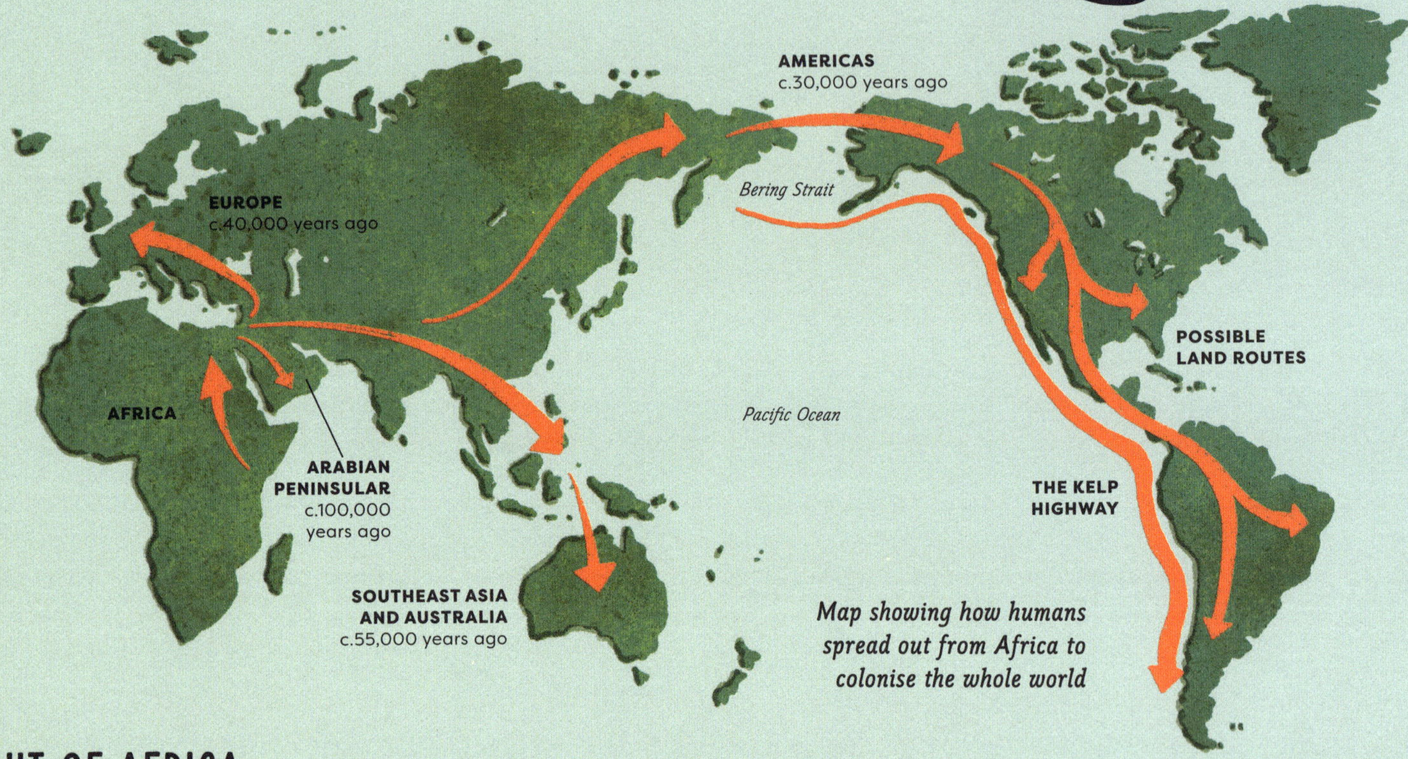

Map showing how humans spread out from Africa to colonise the whole world

OUT OF AFRICA

Early humans were constantly on the move – hunting, gathering wild plants, and travelling great distances when food resources ran out. Our earliest descendants lived in east Africa and then, as populations grew, spread to south and west Africa. From around 100,000 years ago, intrepid explorers left Africa by walking across a narrow bridge of land from the northeast of the continent to the Arabian Peninsula. We cannot be sure why humans made the journey, but sudden shifts in climate may have transformed the desert-like conditions of northeast Africa and the Arabian Peninsula into a lush, green habitat. Large herds of animals may have migrated there, and the people who hunted them followed.

AUSTRALIA BOUND

From the Arabian Peninsula, early humans made their way eastwards, through Asia, reaching Australia by about 55,000 years ago. Much of the journey would have been done on foot as lower sea levels meant that Indonesia was linked to Asia by land, as was New Guinea to Australia. The settlers, who were ancestors of Australia's Aboriginal peoples, at first stayed near the coast and then moved inland. For food, they gathered nuts, fruit and vegetables, and fished and hunted animals, including the now-extinct giant kangaroos and Tasmanian tigers.

FROZEN PLAINS

About 40,000 years ago, early humans crossed into Europe from Asia. Europe was much colder and drier than it is today, and large animals, such as elephant-sized woolly mammoths and rhinoceros, cave bears and sabre-toothed cats, roamed the frozen plains. By 11,500 years ago, many of these large animals were extinct, possibly because they had been overhunted by humans. Humans also settled in Arctic Russia around this time and in areas where winter temperatures could drop to –30°C (–85°F).

YOUNG AMERICANS

Lower sea levels across the globe also exposed a bridge of land, known as the Bering Strait, between northeast Russia and Alaska. From about 30,000 to 12,000 years ago, explorers crossed this land bridge into Alaska and the Yukon in North America. The Bering Strait flooded over again in around 12000 BCE, so settlers continued to spread southwards into North America and down into Central and South America. It was thought these were the ancestors of today's Native Americans who hunted bison, woolly mammoths and other large mammals.

KELP HIGHWAY

Those who reached the Americas by land, however, may have not been the first settlers, as an earlier group of explorers may have reached the continent by sea. Arriving in small boats, they paddled down the Pacific coastline from northeast Asia to North America. The route is known as the 'kelp highway' as it went close to the kelp forests that cover much of the Pacific coastline. Kelp is a type of seaweed that supports a vast ecosystem of fish, shellfish, otters and other sea mammals, providing these brave pioneers with the vital food resources they needed to survive.

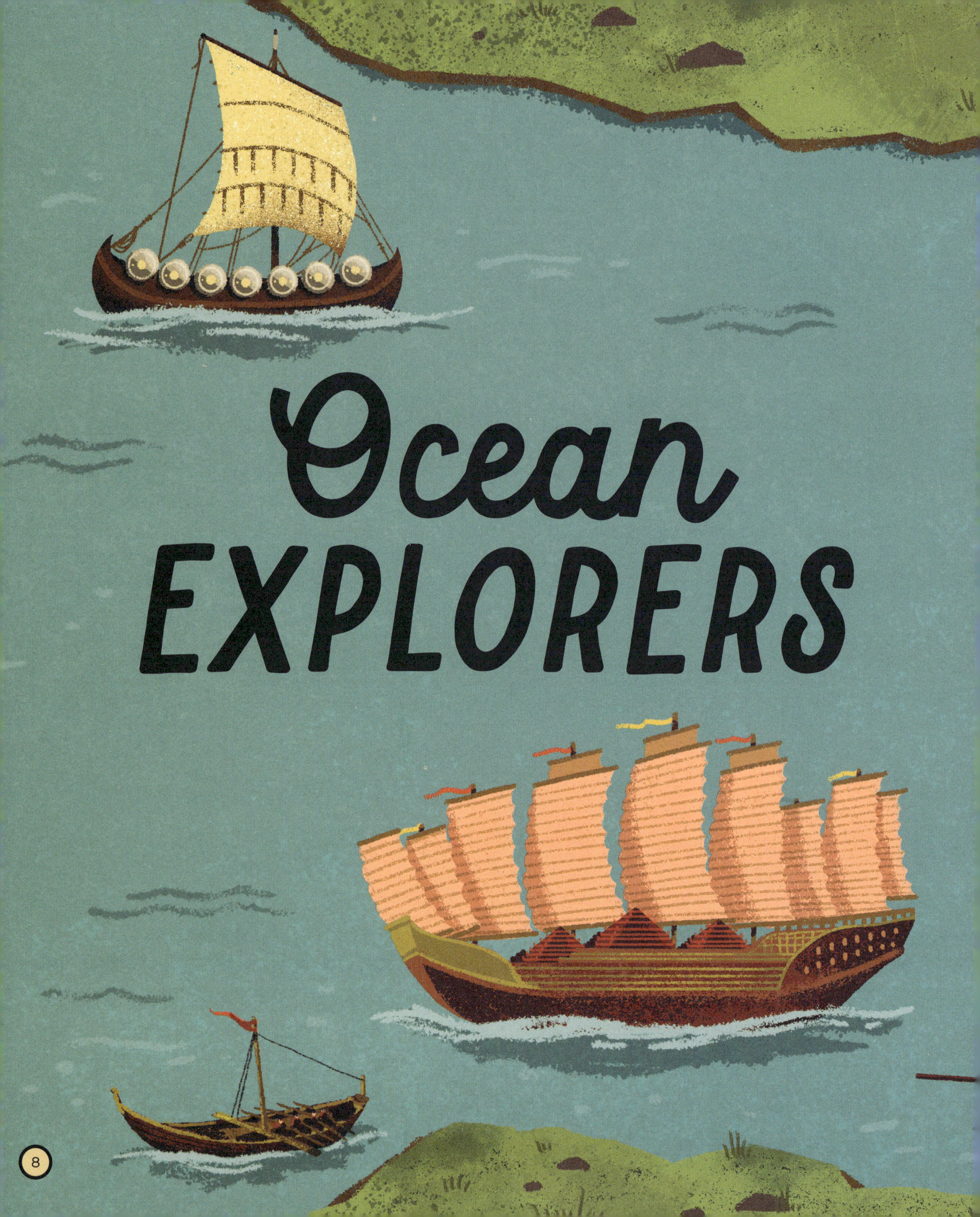

Ocean EXPLORERS

For thousands of years, humans have explored the great seas and oceans. They built boats and eventually sailed all around the world. In doing so, they discovered new lands, spread ideas and mapped the globe for the first time. The very depths of our oceans, a murky underworld that saw the beginning of life on Earth, is yet to be fully explored and may help scientists in their search for life on other planets.

Ancient Seafarers

It was people's curiosity, along with a desire to trade goods, that led to the earliest seafaring adventures – from ancient Egyptian expeditions to the mysterious land of Punt, and journeys by merchants along the Spice Routes, to the voyages of the Phoenicians, who travelled far and wide, spreading ideas and knowledge as they went.

ALONG THE NILE

Some of the first boats ever recorded were built by the ancient Egyptians from around 4000 BCE. Made from papyrus reeds, they sailed up and down the River Nile carrying people and goods. By 2000 BCE, Egyptians were building large wooden cargo ships and had begun venturing into the Mediterranean Sea for trade. In 1493 BCE, the Egyptian pharaoh Queen Hatshepsut organised a famous expedition to the land of Punt (probably located in Somalia). Egyptian inscriptions describe how the ships returned with a huge variety of items including frankincense trees, gold, ebony, monkeys, spices and cosmetics.

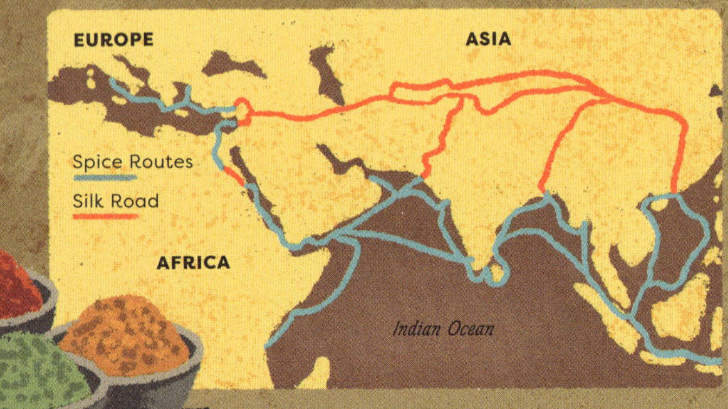

SPICE ROUTES

From our earliest times, people have braved treacherous seas to travel along a network of land and sea routes known as the Silk Road and Spice Routes (named after the goods they traded), which linked Asia in the East with Europe in the West. At first, people probably travelled short distances but the hugely profitable trade in spices and luxury goods prompted ships to sail longer distances.

As early as 2000 BCE, cinnamon was being transported, via many middlemen, from China to the Middle East. Over the next thousand years, ships from the Middle East and India regularly explored the Red Sea and Indian Ocean, their holds stuffed with silver ingots, animal skins, spices and other luxury goods.

THE PHOENICIANS

By 1200 BCE, the most dominant maritime power in the Mediterranean were the Phoenicians. They lived on a coastal strip connecting Asia with Africa. Without enough land to feed their people, they sailed out to sea to find resources and trading partners. In doing so, they became expert sailors and navigators and were known as the 'princes of the sea'. They traded throughout the Mediterranean and may even have travelled as far as Britain and West Africa.

Phoenician trading routes in the Mediterranean

PURPLE PEOPLE

The Phoenicians exported all kinds of products, such as timber, wine, spices, fish, and a very rare and much-prized purple dye made from sea snails, earning them the nickname 'the purple people'. As well as trading goods, they exchanged ideas and knowledge, including the Phoenician alphabet, which is the basis for the Greek and Roman alphabets.

SHIPBUILDERS

To carry their heavy cargoes, the Phoenicians built large ships called biremes that could sail long distances. They had two banks of rowers on either side for greater speed and wide hulls to store goods. The Phoenicians also built war ships equipped with bronze tips so they could ram enemy boats. Many vessels were decorated with carvings or paintings, including eyes to frighten enemies and help their ships 'see'.

With such advanced ships, the Phoenicians were able to establish a vast trade network and settlements along the Mediterranean coastline, laying the foundations for ancient Greece and Rome.

The Phoenicians did not have the compass or other navigational instruments. Instead, they found their way by observing the Sun, stars and natural features.

Polynesia is made up more than 1,000 islands in the middle of the Pacific Ocean – the biggest ocean on the planet. When European explorers arrived there in the 1700s, they were amazed to discover people living on the widely scattered islands. The Polynesians had somehow made their way across vast expanses of open sea – an extraordinary feat of oceanic exploration – without the aid of metal tools, maps, compasses or any modern instruments.

Polynesian Navigators

MASTERY OF THE SEAS

The ocean lay at the heart of Polynesian culture and the survival of its people relied on the mastery of the seas. If islands became crowded, families would set off in search of new lands. The outbreak of disease, war or shortages in food may also have driven Polynesians to explore new islands, which could be thousands of kilometres away.

DOUBLE CANOE

To survive these long voyages, the Polynesians used stone tools to build canoes from wood and braided plant fibres. These included **outrigger canoes**, which have a wooden beam (known as an outrigger) connected to the hull to improve the boat's steadiness on the water. They also sailed in **double canoes**, consisting of two hulls connected by beams. This gave the boat stability and enabled it to carry heavy loads of people, supplies and even animals. A central platform laid over the crossbeams provided more space, and a sail and long oar kept the vessel on course over stormy seas.

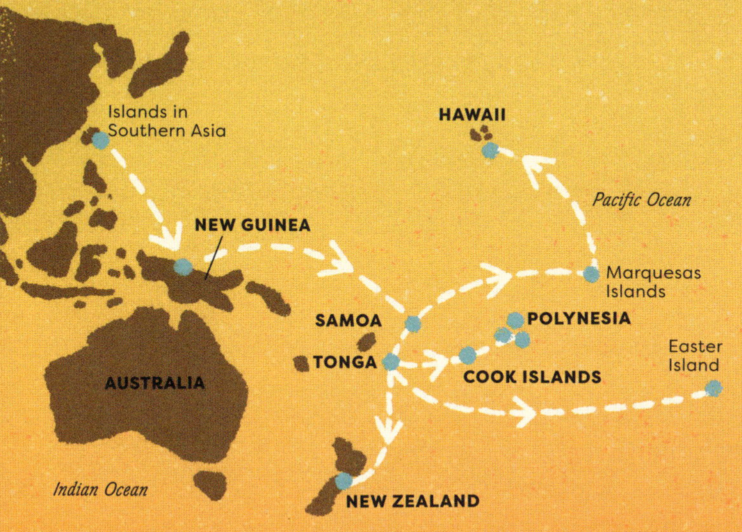

It is thought that the ancestors of the Polynesians came from Southeast Asia, travelling east from New Guinea to the islands of Samoa and Tonga in around 1500 BCE. From there, they fanned out to the Marquesas Islands in around 300 CE and eventually headed to Hawaii in the north, Easter Island in the southeast and finally New Zealand in the southwest by around 1200 CE.

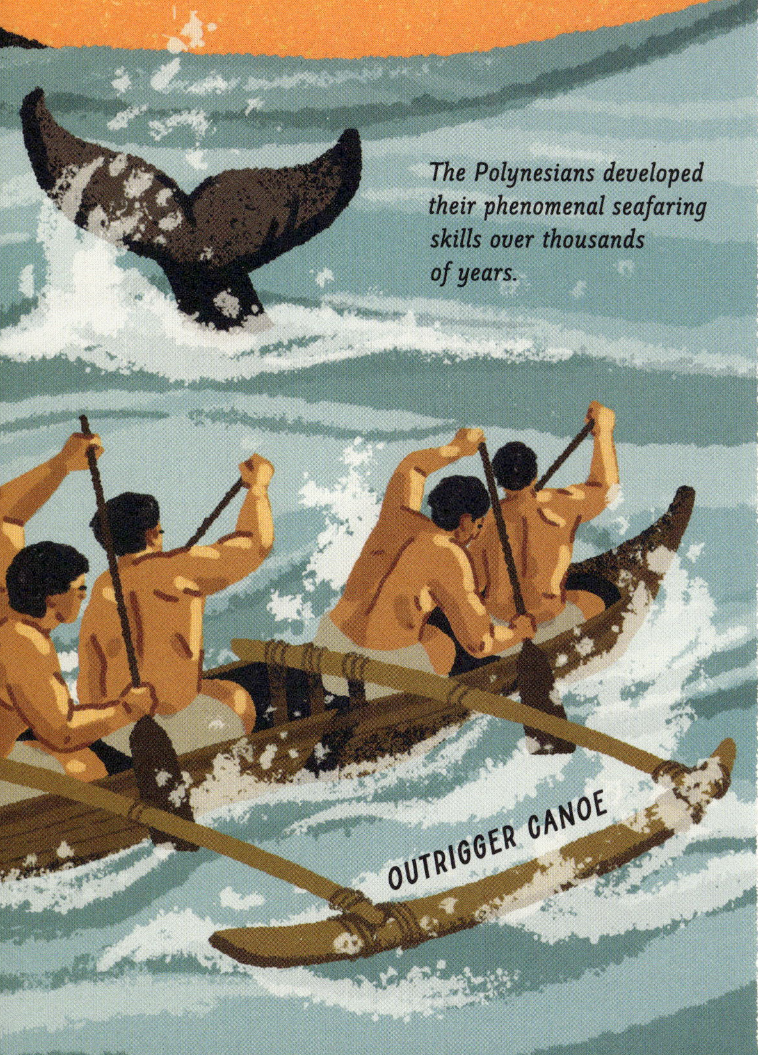

The Polynesians developed their phenomenal seafaring skills over thousands of years.

NAVIGATING BY NATURE

To find their way in the vast ocean, the Polynesians used traditional navigation, or wayfinding, techniques – using the stars, the Sun and other signs of the natural world to guide them.

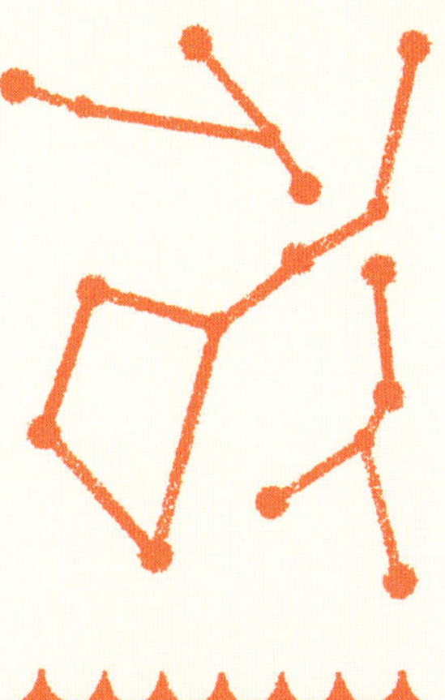

The Sun, Moon and stars

Polynesian seafarers could work out their position and the direction they were sailing by the rising and setting of the Sun, Moon and stars. For each journey, they would memorise a specific sequence of stars that might appear on the horizon.

Waves and currents

Waves and sea currents, which can be detected by the rocking motion of a boat, also helped navigators to know what direction they were heading.

Nearing land

When seeking land, they looked for various natural clues – clouds forming over islands, plant debris drifting in the ocean or the appearance of certain species of fish. Some may have followed the flight path of birds, such as frigate birds, which fly out to sea in the morning and return to land at night.

Migrating animals

Migrating birds or pods of whales may also have been used as a guide to find land. Whales typically give birth near land, and they move slowly enough for a canoe to follow.

Shell maps

As Polynesians had no written language, sailors used charts made of woven plant roots, leaves, sticks and shells to show the direction of wave patterns and position of islands.

Viking Explorers

The Vikings were a seafaring people from Scandinavia who braved open waters in their mighty longships, raiding and settling in much of Europe, and voyaging as far as North America and the Middle East. They were fearsome warriors, but also craftsmen, farmers and traders who travelled in search of new lands and opportunities.

It is thought that Vikings sang chants to help maintain a rhythm when rowing their oars quickly and to keep up morale when sailing in stormy seas.

SURPRISE ATTACKS

From the late 700s CE, Vikings from Scandinavia (mainly areas now known as Denmark, Norway and Sweden) set sail in ships to trade, obtain riches and conquer new lands. Initially, they made hit-and-run attacks on coastal areas in the British Isles and northern Europe. They were fierce fighters, famed for attacking monasteries and plundering their gold, jewels and precious items.

INVASION FORCES

By around 840, the Vikings began to look for new places to live, farm and trade. Eventually, they conquered large parts of England, Ireland and Scotland, and Normandy in France. Vikings from Sweden also crossed the Baltic Sea to Russia. They sailed down rivers as far as the Black Sea, before travelling to Turkey and even Iraq in the Middle East, where they could trade items like fur and wool in return for spices and silver.

LONGSHIPS

To make such epic journeys, Vikings sailed in wooden longships. Powered by sails and oars, these light but strong boats could sail in rough oceans as well as shallow inland waters only 1 m (3 ft) deep. This made the boats perfect for beach landings and surprise attacks.

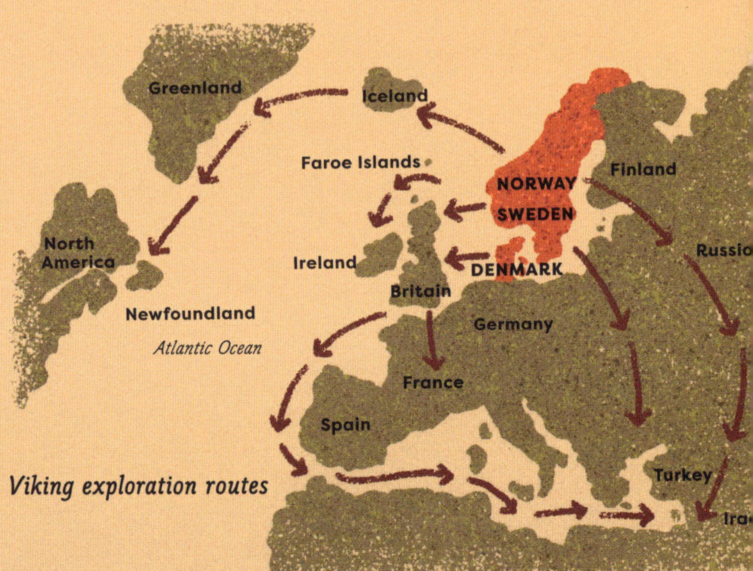

Viking exploration routes

ICY MOUNTAINS

From the 800s, Viking voyagers also began to explore islands in the North Atlantic. A book called the *Landnámabók*, probably compiled in the 12th century, describes how a Viking called Naddodd, who was born in around 830 CE, became perhaps the first person to set eyes on Iceland. Sailing to the Faroe Islands, a storm blew up and pushed him towards a forbidding coast overlooked by ice-topped mountains – the east coast of Iceland. Naddodd went ashore, climbed a mountain but seeing no sign of life, left.

ERIK THE RED

Talk of new lands beyond Iceland enticed Vikings to explore further. According to medieval histories known as the Icelandic sagas, the first Viking to land on Greenland was Erik the Red (named after his red beard and hair, and fiery temper) who had been banished from Iceland in 982 for murder. In the same year, he sailed from Iceland to Greenland.

After spending three years exploring the new land, taking in its icy fjords and lush green valleys, Erik returned to Iceland. It is thought he named the mysterious land Greenland as a way of enticing people to the island. In 985, he returned to Greenland, this time with 35 ships and around 500 people and many animals, including horses, cows and oxen. Stormy weather, however, wrecked some of the ships and forced them to return to Iceland. Only 14 vessels survived the voyage. They went on to establish settlements on the island where Erik had four children, including the famed Viking explorer Leif Erikson (p16).

RAVEN-FLÓKI

In around 870, the Norwegian Flóki Vilgerdarson set out to colonise Iceland, carrying three ravens on board his ship, earning him the nickname 'Raven-Flóki'. By releasing the ravens while at sea and sailing in the direction of one that didn't return, he was able to find Iceland. He and his crew settled on the west coast, and were followed by Ingólfr Anarson, who founded Reykjavík, now Iceland's capital.

VIKING NAVIGATION

To navigate, Vikings used their knowledge of the environment. They observed the Sun and stars to get their bearings and work out their latitude, and looked for landmarks onshore when sailing along the coast. They watched the movements of seabirds, which could tell them the direction of land, and looked for changes in the colour of the sea, and the patterns of waves and clouds. In foggy weather, however, they could easily get lost.

Leif Erikson

(C.970–C.1020)

Around five hundred years before Christopher Columbus stumbled across the Americas, Leif Erikson (nicknamed Leif the Lucky) and a crew of Viking seafarers set sail in search of new lands. Battling the icy winds of the Atlantic Ocean, their wooden ship made it as far as North America. On coming ashore, the ship's crew may have been the first Europeans to walk on American soil.

Leif Erikson arriving at Helluland

WHAT THE SAGAS SAY

Exploration may have run in Leif Erikson's blood, as his father was the fiery-tempered Erik the Red who had founded the Viking settlement of Greenland (page 15). Different tales of Leif's exploits in North America are told in two Viking sagas (stories that were told orally through the generations, until being written down in the 13th century).

According to 'Erik the Red's Saga', Leif discovered North America by accident when he was blown off course while travelling from Greenland to Norway. 'The Saga of the Greenlanders' is thought to be a more reliable source. It recounts how Leif Erikson hears of a 'forested land' that Icelandic trader Bjarni Herjólfsson had spotted from his ship, having been driven there by a storm. Fifteen years later, just after 1000 CE, Leif set off with a crew of 35 to find this mysterious land.

LANDS OF FLAT ROCKS AND FORESTS

After crossing the Atlantic, they reached a barren, flat land with icy mountains, which they named Helluland ('land of flat rocks'), a location that may have been Baffin Island in present-day Canada. They sailed further south and arrived at a forested land with white, sandy shores, which Leif named Markland ('land of forests'), which is likely to have been Labrador, Canada.

LAND OF GRAPEVINES

They eventually set up camp, probably at L'Anse aux Meadows on the northern tip of modern-day Newfoundland, Canada, where a Viking settlement has been found. They spent a winter there and explored the surrounding area where they found lush meadows, rivers teeming with fish, grapes and much-needed timber, which they brought back to Greenland. They named the new land Vinland ('land of grapevines'). On his return to Greenland, Leif took over from his father as chief of Greenland and never returned to Vinland.

ARCHAEOLOGICAL EVIDENCE

For many years it wasn't known if the stories told in the sagas were true, but in the 1960s, archaeological excavations in Newfoundland, Canada, uncovered evidence of Viking dwellings, proving that they really had been there.

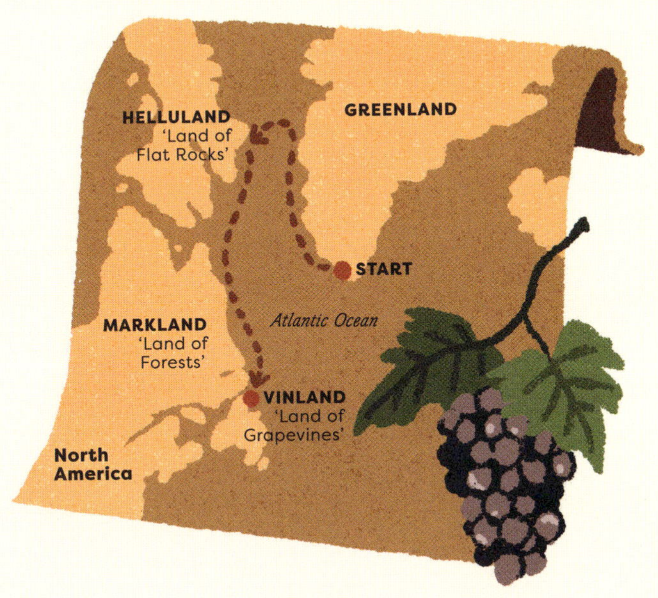

> **'After being tossed about at sea for a long time, [Leif] chanced upon land where he had not expected any to be found. Fields of self-sown wheat and vines were growing there.'**
>
> 'The Saga of the Greenlanders'

GUDRID THORBJARNARDÓTTIR

Another Viking explorer described in the sagas was a woman named Gudrid Thorbjarnardóttir, known as 'Gudrid the Far-Traveller' (980–1019). So the stories go, she journeyed to Norway and Greenland, and later, set off with her husband on an expedition to the strange new world of Vinland. There, she gave birth to a son, Snorri, the first baby born to a European in North America.

According to the sagas, Gudrid lived in Vinland for three years, then returned to Iceland when the Viking settlement in North America was abandoned, possibly because of hostilities with local people. To survive such an arduous trip, give birth to a healthy child and face all the hazards of a new settlement is an impressive achievement and showed that Viking women could be as brave and adventurous as Viking men. She went on to run a farm in Iceland and travelled to Rome in her later years.

Zheng He

(1371–1433)

›››››››››››››››››››››››››

In the early 1400s, the Chinese explorer and admiral Zheng He commanded a huge fleet of ships on daring expeditions across the Indian Ocean as far as Arabia and the east coast of Africa. His aim was to show the power and glory of China and to secure sea routes for trade. His seven expeditions travelled further than any in Chinese history, his crew of thousands sailing in the biggest ships the world had ever seen.

HUMBLE BEGINNINGS

Zheng He was born in the poor Chinese province of Yunnan in 1371, just as the new Ming Dynasty was taking over China. At the age of around ten, he became a servant of the Ming emperor's son Zhu Di. He went on to be a skilled soldier in the Ming army, and in 1402, helped Zhu Di take the throne and proclaim himself Emperor Yongle (meaning 'perpetual happiness').

Yongle was an ambitious ruler and was determined to extend Chinese power and control trade in the Indian Ocean. With Zheng He as admiral, he set about building a huge fleet of ships.

TREASURE SHIPS

In 1405, Zheng He set off on his first voyage. Legend has it that the fleet included 62 enormous 'treasure ships' 122 m (400 ft) long and 50 m (160 ft) wide – the size of a football pitch and five times bigger than the ships sailed by European explorers later in the century. Historians think the ships could not have been more than half that size, but that was still very impressive for the time.

The biggest vessels had four decks and watertight compartments, making the boats more stable and enabling them to carry huge amounts of drinking water.

It was said that the ships were a dazzling sight with red sails, yellow flags fluttering on their rigging and huge birds painted on their hulls.

The magnetic compass

MAGNETIC COMPASSES

Zheng He's ships were equipped with paper charts, maps and magnetic compasses. Historians think the Chinese were the first to use the magnetic compass for navigation as early as the 11th century. News of the invention spread fast – English sailors are first recorded using it in around 1187.

SEVEN VOYAGES

The first voyage sailed to Vietnam and Thailand, then all the way to India and Sri Lanka. Zheng He continued to sail on his voyages for the next 28 years, travelling westwards across the Indian Ocean to Arabia and along the east coast of Africa to Somalia and Kenya. He set off on his seventh and final voyage in 1431 but died in 1433 before the fleet arrived back in China.

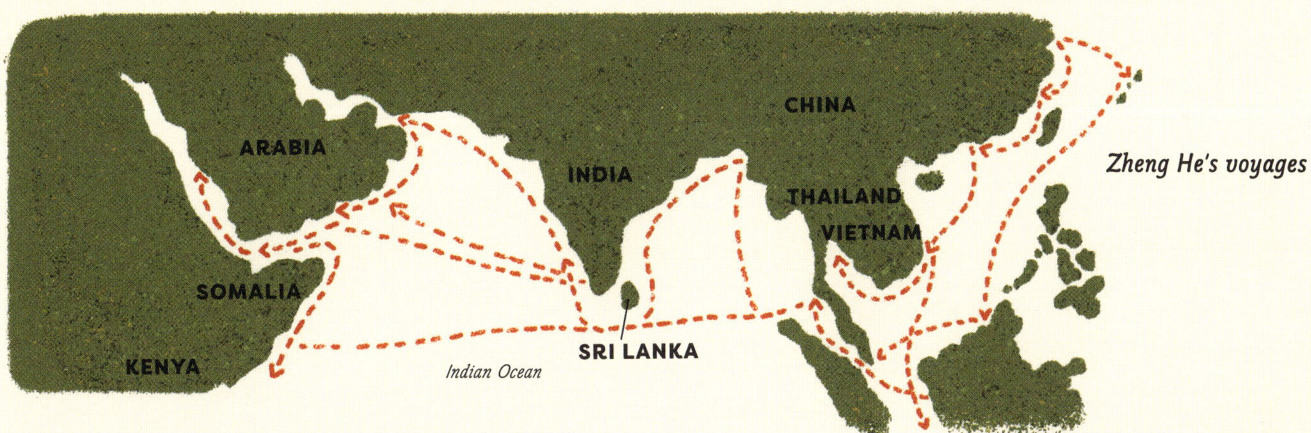

Zheng He's voyages

ELEPHANTS, OSTRICHES AND GIRAFFES

During his voyages, Zheng He visited more than 25 countries, many of which had never had any contact with China before. He traded goods, battled pirates and experienced stormy seas and shipwrecks.

He also brought back diplomats from various countries to meet with the Chinese emperor, and all kinds of items, including spices, precious stones and even exotic animals such as ostriches, rhinos, zebras, lions and giraffes. These animals were met with great wonder in China, particularly the giraffe, which was entirely unknown in the East and thought to be a mythical creature, similar to a unicorn. The giraffe eventually became an emblem of Emperor Yongle's reign and the remarkable overseas trade missions.

'We have traversed more than 100,000 li* of immense water spaces and have beheld in the ocean huge waves like mountains rising in the sky, and we have set eyes on barbarian regions far away.'

From a 15-century inscription said to be by Zheng He

*50,000 km (30,000 miles)

The Age of Discovery

VOYAGES OF DISCOVERY (1487–1771)

Bartolomeu Dias (1487–1488)

John Cabot (1497–1498)

Vasco da Gama (1497–1499)

Jacques Cartier (1534–1542)

Christopher Columbus (1492–1504)

James Cook (1768–1771)

In the late 1400s, European sailors began to explore the oceans in search of new trade routes and vast undiscovered lands. They sailed around the globe and set up trading posts and colonies in Asia, Africa and the Americas. Knowledge about the world grew and European nations amassed great riches, but native peoples died in their millions and a brutal system of slavery emerged.

ROUTE TO RICHES

European nations were eager to find new sea routes to trading partners in the East so they could bring back valuable cargoes of spices, silks, gold and silver and other luxury goods. Travelling by land to Asia was slow and dangerous, and trade with the East was controlled by Muslim middlemen in the Middle East. So, the hunt was on to find quicker routes by sea. Improvements in navigation and shipbuilding also enabled seafarers to travel further without keeping close to shorelines.

DIAS AND DA GAMA

From Europe, the Portuguese explored the oceans in fast sailing ships, travelling along the west coast of Africa in search of gold and a route to the East. In 1488, Portuguese nobleman Bartolomeu Dias reached the southern tip of Africa, the Cape of Good Hope, paving the way for Vasco da Gama to cross the Indian Ocean to the east (p26). He and other Europeans often enlisted the help of Indian and Arab seafarers who were skilled in navigating across the Indian Ocean and the Red Sea.

CHRISTOPHER COLUMBUS

Spain also dreamt of new sea routes and riches, which they found in the Americas thanks to Italian explorer Christopher Columbus (p24). In the search for Asia, Columbus stumbled across the Caribbean, opening the door to centuries of European colonisation of the American continent.

CABOT, CARTIER AND COOK

In 1497, the Italian explorer John Cabot sailed an English ship west across the Atlantic and sighted land off the east coast of Canada – he called it 'New-found-land', as it is still named today.

John Cabot

From 1534, Frenchman Jacques Cartier explored the Atlantic coast of Canada. Cabot and Cartier's voyages would lead to English and French settlements in North America. In the 1700s, English navigator James Cook (p34) sailed the South Pacific and mapped New Zealand and the east coast of Australia during his scientific voyages of discovery.

A NEW WORLD

The Age of Discovery created many new opportunities for trade around the world and brought great wealth to European colonisers. The discovery of new lands and continents also rapidly advanced European knowledge of the world. Navigation improved and nautical maps depicted not just land but also sea routes and currents. A whole new world of plants and animals was also introduced to Europeans. Only in the Americas could you see a llama, and European diets began to include such delicacies as maize, potatoes, tomatoes, cocoa and chillies.

DEADLY LEGACY

In reaching new continents, however, Europeans brought with them new diseases, such as smallpox and influenza, which decimated native populations in the Americas and Australasia. Meanwhile, enslaved people from Africa (who brought diseases like yellow fever with them) were taken to work on European plantations and mines in the Americas. Chained and herded into the holds of ships, millions died en route or from overwork or disease once they reached their destination. From 1502–1867, up to 14 million Africans were transported across the Atlantic – the biggest forced migration of humans in history.

Mapping the World

The Age of Discovery led to a better understanding of the globe's great continents and oceans, and the first true maps of the world. With better maps and navigational tools, seafarers could plan routes and determine with more certainty what lay beyond the horizon.

EARLY MAPS

Prior to the Age of Discovery, maps of the world were limited to what mapmakers knew about the globe. Ptolemy's world map, created in 150 CE in Roman Egypt, influenced world maps for centuries but it featured only Europe, Asia and Libya in Africa. The most advanced medieval world map was created by Arab geographer Muhammad al-Idrīsī in 1154. It combined knowledge from Arab merchants and explorers who had voyaged to the Indian Ocean, Africa and Asia. Large parts of the world, however, were missing, including most of Africa, the Americas and other lands that were still unknown in the Old World.

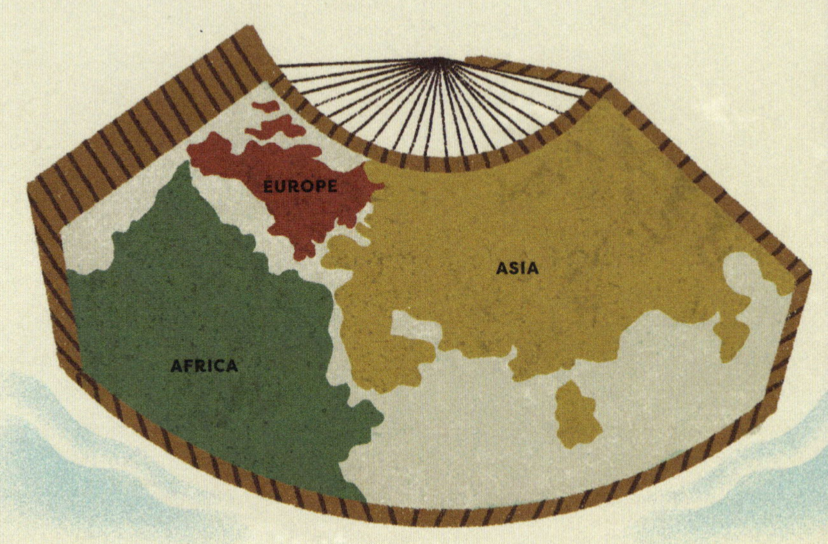

1517 map of the world inspired by Ptolemy. The map extends only so far south as the Equator. Ptolemy believed that the land south of this was inhabited by monsters.

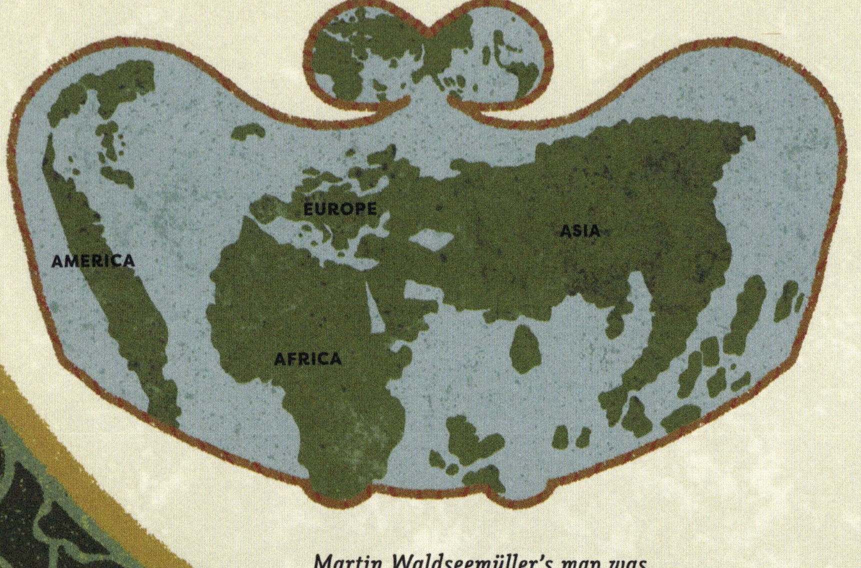

Martin Waldseemüller's map was the first to use the name 'America'.

NEW DISCOVERIES

With the discovery of new continents and oceans, world maps became more accurate. In 1507, America was shown for the first time as a distinct region separate to Asia on a map created by German mapmaker Martin Waldseemüller. Meanwhile, the vast size of the Pacific Ocean, as revealed during Ferdinand Magellan's circumnavigation of the world (1519–1522), was mapped with more accuracy by Portuguese cartographer Diogo Ribeiro in 1527.

MAP STYLING

In 1569, Flemish mapmaker Gerardus Mercator created a world map that helped sailors sail around the world. Its grid framework, on which the spherical planet is flattened into a two-dimensional map, enabled seafarers to chart a course in straight lines and still helps navigators today. In 1570, Mercator encouraged Flemish cartographer Abraham Ortellius to create the world's first modern book of maps, called an atlas. Entitled *Theatrum Orbis Terrarum*, the atlas combined everything Europeans had learnt about the world in one book.

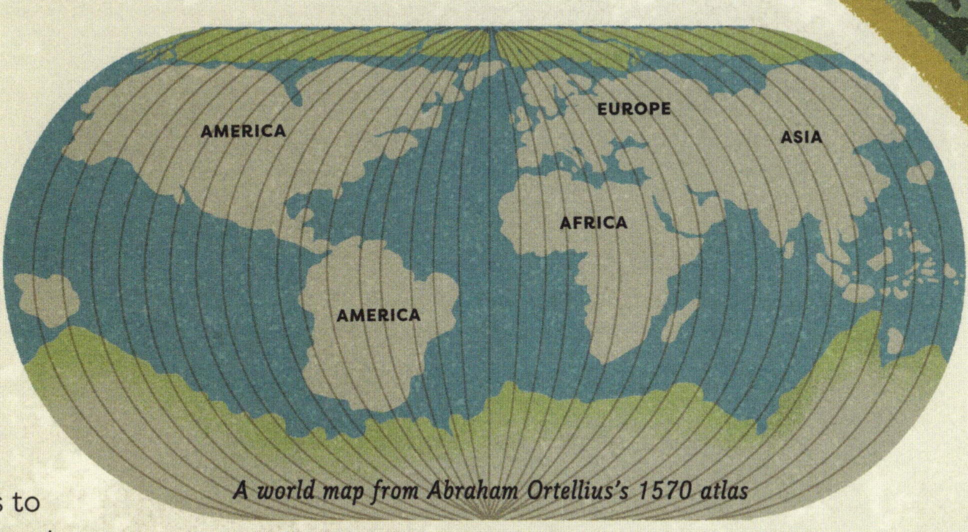

A world map from Abraham Ortellius's 1570 atlas

You can't use an old map to explore a new world.

Albert Einstein

NAVIGATIONAL TOOLS

During the Age of Discovery, sailors continued to use medieval navigational tools such as the compass (for determining direction) and astrolabe (a tool that measures the distance of the Sun and stars above the horizon) but replaced them with more effective versions. These tools helped seafarers determine lines of latitude, a series of lines depicted on maps that mark the distance north or south of the Equator (the central point between northern and southern hemispheres).

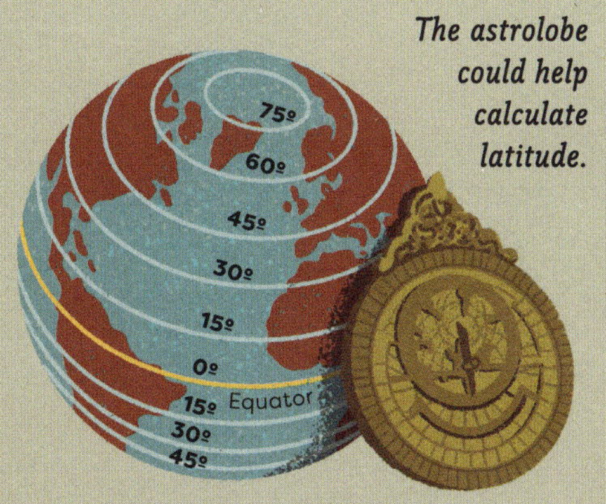

The astrolobe could help calculate latitude.

A chronometer for measuring longitude

Longitude, which marks the distance east and west of the Greenwich meridian, is much more difficult to determine. An accurate clock is required to measure longitude properly, which was impossible to do on rolling seas with pendulum clocks. As sea captains couldn't measure their location east or west, they often miscalculated their position and their ships ran aground or sunk. The problem was finally solved in the 1750s, when English clockmaker John Harrison constructed a chronometer that kept nearly perfect time on board ships. Sea captains could finally navigate the seas with real accuracy.

Christopher Columbus

(1451–1506)

Christopher Columbus is famed as the explorer who discovered the Americas. But actually, an estimated 100 million people were living there already, and we now know that the Vikings had briefly reached the Americas hundreds of years before Columbus. His voyages, however, would bring great riches to Spain and began the European settlement of the Americas. He also exploited the native people, a pattern that would be repeated by Europeans across the continent.

AN ATLANTIC ADVENTURE

Italian navigator Christopher Columbus was an adventurous spirit who, like many Europeans, was keen to find a sea route to Asia. He convinced the Spanish king and queen to finance the trip so that he might find new lands, convert people to Christianity, discover gold and other riches, and secure a supply of spices for Spain.

He set off from Spain on 3 August 1492 with three ships, the *Niña*, *Pinta* and *Santa Maria*. Instead of travelling east around Africa, Columbus sailed west across the Atlantic Ocean. Others had tried and failed to cross the Atlantic, but Columbus, who was a skilled and experienced seafarer, knew that if he sailed south to the Canary Islands first, more favourable winds would then help the fleet across the rough waters of the Atlantic. He also knew that if he failed to find land, he could head for home at any time by sailing north in the band of east-blowing winds.

To the day he died in 1506, Columbus remained convinced he had reached the East Indies.

LAND IS SPOTTED

Five weeks later, land was spotted, and the ships landed on a small island in the Bahamas. Columbus then explored Cuba and Hispaniola (now Haiti and the Dominican Republic). Convinced he had reached the East Indies in Asia, he named the islands the 'Indies' (now called the West Indies). On meeting the Arawak people who lived on these Caribbean islands, Columbus wrote that they could be easily conquered and converted to Christianity. And, noticing that they wore gold ornaments, he took some of them prisoners and insisted they take him to the source of their gold.

Voyages of Christopher Columbus

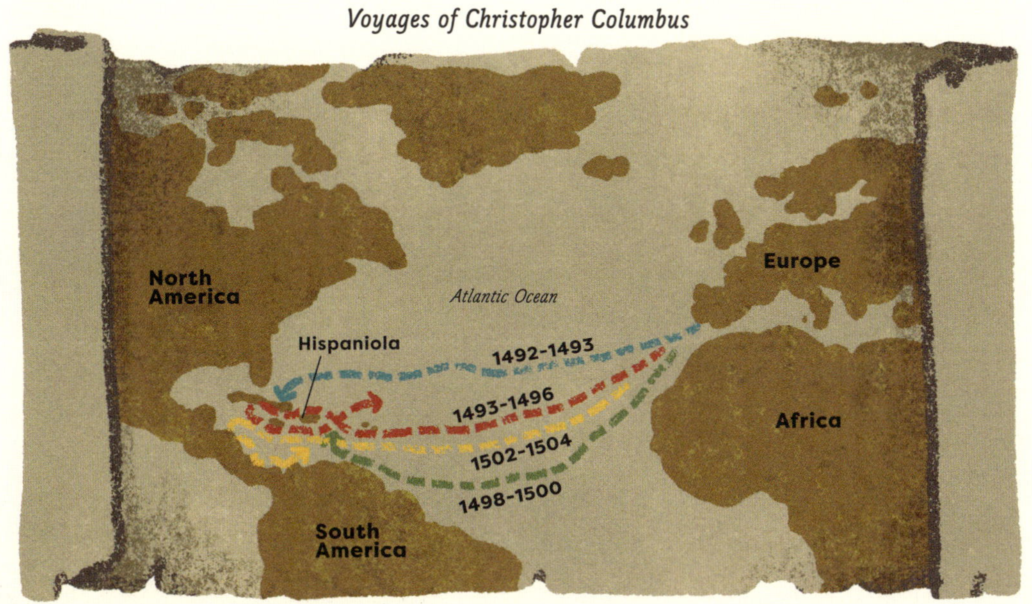

On Christmas Day 1492, the *Santa Maria* hit rocks and Columbus was forced to return on the *Niña* and *Pinta*, leaving behind some of his men to set up a settlement in Hispaniola. After battling through terrible storms, the two ships arrived in Spain in March 1493.

A BRUTAL LEGACY

Columbus made another three voyages across the Atlantic, landing in Central America and Venezuela in South America, and bringing with him more Europeans. However, to enforce Spanish rule, Columbus was known to treat native people brutally: he forced the Taino people of Hispaniola to look for gold and work on plantations, and he sent at least 500 slaves back to Spain.

European diseases also spread through the islands, with the population of Hispaniola plummeting from around 250,000 in 1492 to 14,000 by 1517. Columbus's voyages paved the way for centuries of European exploration and colonisation of the Americas. They also led to the exploitation of Native American populations and the devastation of their way of life.

Columbus brought back pineapples, tomatoes, turkeys and parrots.

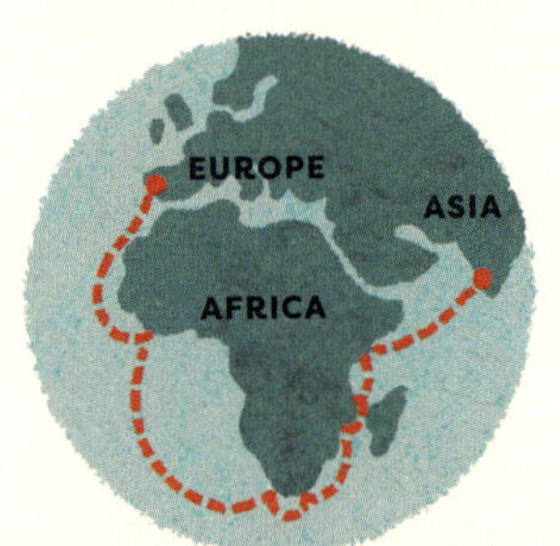

Vasco da Gama

(C.1460–1524)

Portuguese sailor Vasco da Gama was one of the most well-known explorers of the Age of Discovery. He was the first European to reach India by sea on a voyage that would finally link Europe with the East. On his return, he was celebrated as a hero, although in the East he was known for his violent attacks of Arab ships and traders.

IN SEARCH OF SPICES

In July 1497, Vasco da Gama left Portugal with a fleet of four ships. His mission was to find a sea route to India and to break into the lucrative spice trade. Sailing down the west coast of Africa, he made it round the southern tip, stopped several times on the east African coast and headed across the Indian Ocean, finally dropping anchor at Calicut in India on 20 May 1498. He had succeeded where many explorers before him had failed!

SUGAR AND HONEY

On his arrival, da Gama presented the ruler of Calicut with gifts, which included 'four scarlet hoods, six hats, four strings of coral, a chest of sugar and a cask of honey'. The ruler, who expected gold from visitors, was insulted by the gifts and had no interest in making a trade agreement with the Portuguese. The fleet's return trip across the Indian Ocean, battling against monsoon winds, took a punishing 132 days. By the time they reached Portugal, only 54 of the original crew of 170 had survived the 39,000-km (24,000-mile) journey, then the longest ocean voyage ever made.

VIOLENT ATTACKS

During later trips to India, da Gama attacked Muslim ports, set fire to a ship full of Muslim pilgrims, killing all 400 passengers, and forced the ruler of Calicut to sign an agreement. He was eventually made Viceroy of India, where he died of Malaria in 1524. His voyages and sometimes his violent methods were repeated by future Portuguese explorers who established trading posts on the east coast of Africa and across Asia.

Ahmad Ibn Mājid

(C.1432–C.1500)

>>>>>>>>>>>>>>>>>>>>>>>>>

At the time of Vasco da Gama's voyage, the Arab world understood the Indian Ocean far better than any European. Ahmad Ibn Mājid was an Arab navigator whose seafaring skills were known across the Indian Ocean, earning him the nickname 'Lion of the Sea'. His books, charts and maps were used for many years and helped transform how sailors navigated the ocean.

NAVIGATOR AND WRITER

Born in Julfar (in present-day United Arab Emirates), Ibn Mājid came from a great ocean-going family and by the age of 17, he could navigate a ship by himself. He went on to write many books about navigation, much of it based on knowledge that had been passed on by generations of Arab sailors. He wrote about everything from monsoon winds, tides and currents to the movement of the Sun and Moon and how to navigate using the Pole Star (a bright star that can be seen in the sky in northern parts of the world when you look north).

MASTER SAILORS

Up until the 1500s, Arab merchants regularly sailed the sea routes of the Red Sea, the Indian Ocean and along the east coast of Africa, their Arabian dhows filled with cardamon, ginger and other prized goods. Ideas and techniques of seafaring were also exchanged along these routes.

It was once thought that Ibn Mājid helped Vasco da Gama sail to India. Recent research suggests this was unlikely and that da Gama instead enlisted the help of an Indian sailor. However, the Portuguese may well have relied upon da Gama's books and the collective knowledge built up by Arab sailors across the Indian Ocean.

'They [the Portuguese] admit we have a better knowledge of the sea and navigation and the wisdom of the stars.'

Ahmad Ibn Mājid

Ginger and cardamom

Ferdinand Magellan

(1480–1521)

Portuguese explorer Ferdinand Magellan was one of the great explorers of his time. He was the first European to cross the Pacific Ocean and he organised the expedition that would be the first to circumnavigate (travel around) the Earth. The voyage was a treacherous one, and many of the 270-strong crew died of starvation or disease. Magellan himself was killed and never made it home, but some of his crew did. After a gruelling three-year journey, the last remaining ship, captained by Juan Sebastián Elcano, finally arrived back in Spain.

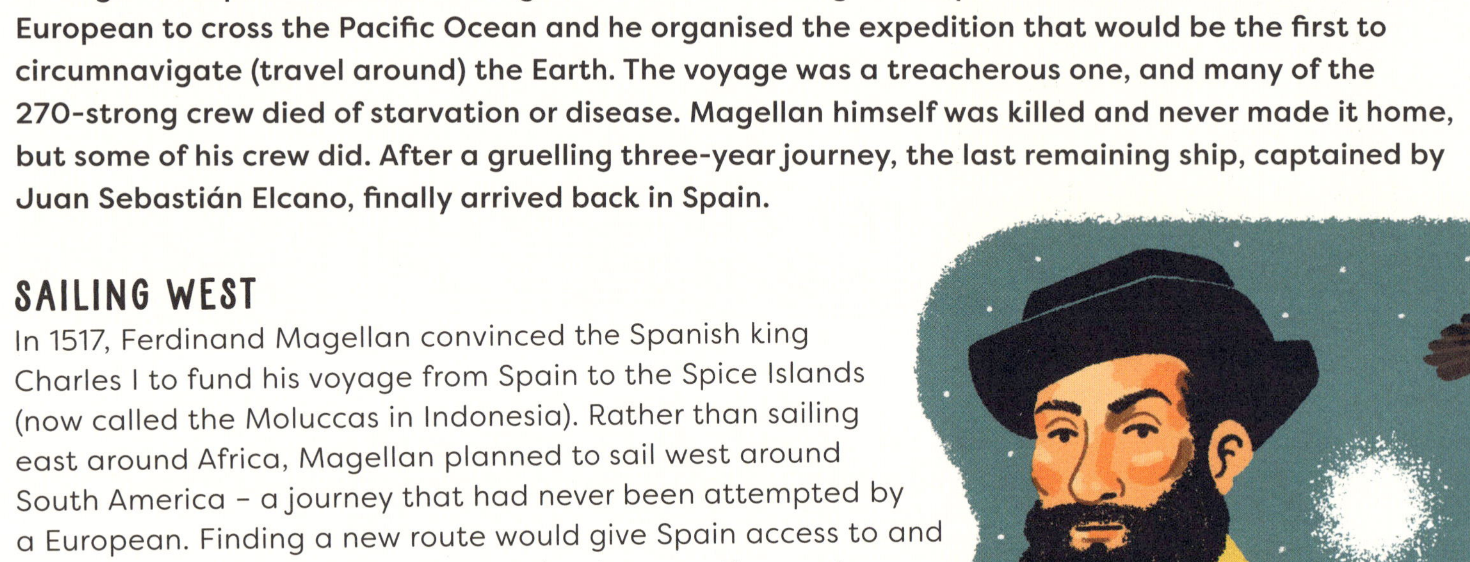

SAILING WEST

In 1517, Ferdinand Magellan convinced the Spanish king Charles I to fund his voyage from Spain to the Spice Islands (now called the Moluccas in Indonesia). Rather than sailing east around Africa, Magellan planned to sail west around South America – a journey that had never been attempted by a European. Finding a new route would give Spain access to and control of the spice trade, which at the time was the most lucrative in the world (p10).

On 20 September 1519, Magellan's fleet of five ships left Spain and headed out into the Atlantic Ocean. The fleet sailed to Brazil and then along the coast of South America to Patagonia. There, Magellan put down a serious mutiny (rebellion) led by some of his crew, many of whom were Spanish and distrusted their Portuguese captain.

CROSSING THE MIGHTY PACIFIC

Continuing south, Magellan's fleet rounded the southern tip of South America and entered a dangerous, windswept channel, now called the Strait of Magellan. For over a month, they battled their way through until they finally emerged into open sea on 28 November. They had made it through to the Pacific Ocean – a momentous achievement.

As no European had sailed across the Pacific Ocean, Magellan had little idea of its vast size. He thought it would take four days to find the Moluccas, but instead the fleet spent nearly four months in search of land. Food and water ran dangerously low. The crew was forced to eat sawdust, rats and old biscuits full of worms and stinking of rats' urine, and scurvy caused the men's gums to swell up.

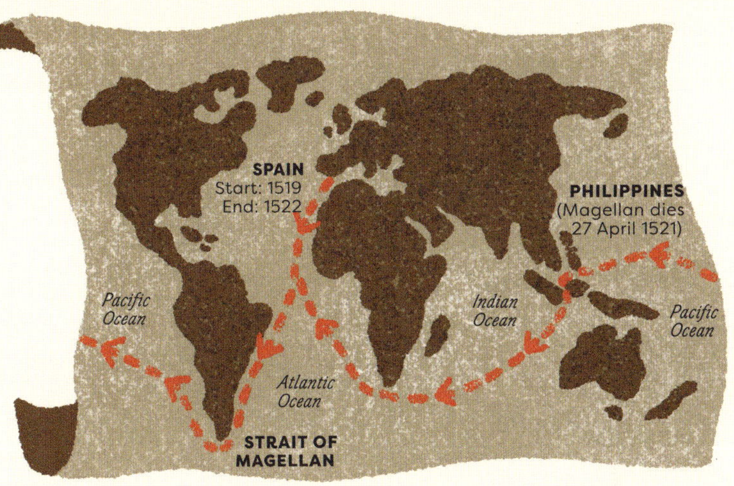

THE FATE OF MAGELLAN

In March 1521, the fleet, now made up of three ships, sailed on to the Philippines. They planned to take possession of the islands for Spain but on 27 April 1521, Magellan was killed in battle.

> **'The sea is dangerous and its storms terrible, but these obstacles have never been sufficient reasons to stay ashore.'**
>
> Ferdinand Magellan

JUAN SEBASTIÁN ELCANO

After Magellan's death, the Spanish captain Juan Sebastián Elcano took charge of the fleet and continued on to the Spice Islands. In December 1521, the last remaining ship, the *Victoria*, set sail for Spain, its hold filled with spices. Elcano and his 17 men finally made it back to Spain on 6 September 1522. Not only were they the first explorers to circumnavigate the globe, as the fleet had not 'fallen off' the end of the world as some predicted, their epic 69,000-km (43,000-mile) voyage also proved that the world was round and not flat.

The Emperor Charles V recognised Elcano's epic achievement by awarding him a coat of arms featuring a globe with the words *Primus circumdedisti me* (You went around me first).

Hasekura Rokuemon Tsunenaga

(1571–1622)

In the early 1600s, Japanese seafarers were also exploring the oceans. Over seven years, Hasekura Tsunenaga crossed the Pacific to the Americas, travelled overland through Mexico, then sailed all the way to Europe where he dazzled the likes of the king of Spain and the pope in Rome. Yet until the 19th century, the story of Hasekura was virtually unknown in Japan.

ADVENTURES IN THE AMERICAS

Hasekura was a samurai – a member of the ruling military class – who had been ordered by a regional lord to lead a voyage to Europe. His mission was to establish trade links with the Spanish crown and to meet with the pope in Rome.

On 28 October 1613, he set sail from Japan in a newly made ship, with hundreds of people on board. After a gruelling two-month voyage across the Pacific, they landed in present-day California, USA, and sailed down the coastline to Acapulco before travelling overland across Mexico (then part of the Spanish Empire). After a six-month-stay, Hasekura sailed to Cuba, then in July 1615, left for Europe – probably the first Japanese to sail across the Atlantic.

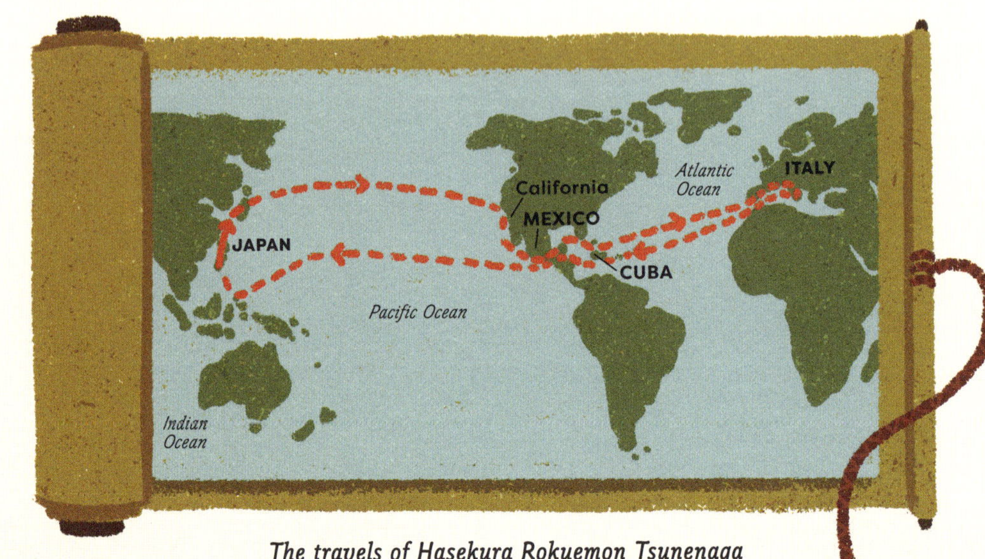

The travels of Hasekura Rokuemon Tsunenaga

SPANISH RECEPTION

Three months later, the Japanese party arrived in Spain, where a lavish welcome awaited them. Eyewitnesses commented on the great spectacle of the visitors from the East, many of whom were dressed in brightly coloured embroidered silks with swords tied at their waists.

In January 1615, they met with the Spanish king in Madrid and, shortly after, Hasekura was baptized into the Catholic Church and renamed Felipe Francisco Faxicura. After travelling around Spain, the Japanese party sailed across the Mediterranean in three Spanish ships towards Italy; however stormy weather forced them to drop anchor at the French harbour of Saint Tropez.

> 'They blow their noses in soft silky papers the size of a hand, which they never use twice, so that they throw them on the ground after usage, and they were delighted to see our people around them precipitate themselves to pick them up . . . Their swords cut so well that they can cut a soft paper just by putting it on the edge and by blowing on it.'
>
> Account of the Japanese party as they travelled through France to Italy

WHEN IN ROME

Arriving in Italy, Hasekura headed to Rome and was escorted to the pope's residence, the Vatican, on horseback. There, he handed the pope two gilded letters, one promising that his lord would spread Christianity in Japan and the other requesting missionaries. In exchange, the pope granted Hasekura Roman citizenship, an honour awarded only to the very few.

SAKOKU

The Japanese party then returned to Spain, Mexico and the Spanish colony of the Philippines, eventually reaching Japan in September 1620. Hasekura's trip, however, was badly timed, as Japan had begun to persecute Christians and was about to retreat into *sakoku*, a policy of isolation that saw the country seal itself shut for nearly two centuries and end trade between Japan and Europe.

Hasekura had sailed thousands of kilometres across oceans, charming European leaders as he travelled, only to return to a nation that had decided to cut itself off from the rest of the world. Hasekura was no doubt aware that as a converted Catholic he faced great danger, but he when died in 1622, it was probably from natural causes.

Jeanne Baret

(1740–1807)

In 1766, Jeanne Baret set off on a perilous journey across the Pacific Ocean to collect new and exotic plants. As women were barred from French naval vessels, she was forced to disguise herself as a man, placing her in extreme danger. She went on to discover many new species and became the first woman known to sail all around the world.

SECRET PLAN

In 1766, the French government asked the admiral and explorer Louis-Antoine de Bougainville to captain an expedition to discover new territories for France. He asked the young scientist Philibert Commerson to join the party to collect plant species. Commerson agreed to go, but he wanted to bring with him his assistant and former housekeeper Jeanne Baret.

Despite having little formal education, Baret had acquired an excellent knowledge of plants. However, the French navy did not allow women on board ships, so she and Commerson hatched a secret plan. Jeanne dressed as a man and gave herself the name of 'Jean' (French for John), and the two pretended not to know each other.

They boarded the ship *Etoile* on 1 February 1766 and sailed southwest towards South America. Fortunately, Commerson had been given use of the captain's cabin and its private toilet, meaning Baret wouldn't have to share the communal 'heads' (toilets) used by the all-male crew.

SUSPICIOUS MINDS

Despite Jeanne's efforts, Bougainville's diary shows that the crew soon became suspicious.

> 'For some time, a rumour had been circulating . . . that Mr De Commercon's servant, named Baré, was a woman. His structure, his caution in never changing his clothes or carrying out any natural function in the presence of anyone, the sound of his voice, his beardless chin, and several other indications had given rise to this suspicion . . . '
>
> Bougainville's Journal

SEEING RED

In mid-June, the ship reached Brazil, and Baret and Commerson went ashore to search for plants. They found a beautiful red flower, which they named Bougainvillea after the expedition's captain. At the time, Commerson was suffering from painful leg ulcers, so it's likely that Baret found this and many of the plants during the expedition.

Continuing south, the ship stopped in Uruguay, where the pair looked for cactus plants, home to cochineal beetles. When crushed, the insects produce a red dye that was highly prized in Europe. The ship then sailed around the southern tip of South America, where Commerson spotted a new species of dolphin (now named Commerson's dolphin), along with penguins and enormous elephant seals.

Jeanne Baret's voyage

RATS AND BOILED LEATHER

Sailing on into the Pacific, the explorers stopped at Tahiti, where it is thought that Baret's true identity as a woman was discovered. By this time, conditions on board the ship were dire, with the crew reduced to eating rats and gnawing on boiled leather. Continuing westwards, the ship struck a jagged wall of coral rock, the Great Barrier Reef, before strong sea currents and squally winds forced them to land on New Ireland, in Papua New Guinea.

BRAVERY

In 1768, the ship reached Mauritius, where Commerson and Baret stayed to collect plants. It was here that Commerson died in 1773. Baret returned to France a year or two later and so completed her circumnavigation of the globe. The two botanists had collected more than 6,000 plants and identified many new species, much of which was undertaken by Baret. Her bravery under such extreme and perilous conditions was extraordinary.

Captain James Cook

(1728–1779)

Between 1768 and 1779, Captain James Cook led three voyages to the Pacific Ocean, exploring and mapping the Pacific islands, New Zealand, Australia, the Antarctic and many other areas previously unknown to Europeans. These expeditions greatly increased Europe's knowledge of the world. However, they opened the way to the colonisation of Australia and New Zealand and the dispossession of their indigenous peoples.

SECRET MISSION

In 1768, navigator and cartographer James Cook was chosen to command a British scientific expedition on HMS *Endeavour*. The mission was to observe the rare astronomical event of the planet Venus passing in front of the Sun. But Cook had another, secret mission – to see if he could find a mysterious southern continent which was believed to encircle the South Pole. Europeans had been hunting for it for over two centuries and Britain wanted to lay claim to it.

THE FIRST VOYAGE

Setting sail in 1768, the *Endeavour* arrived in Tahiti in April 1769, where the crew observed the transit of Venus. They also admired the islanders' dyed skin, which led to the fashion of tattooing among the sailors. They continued on to New Zealand, where Cook mapped much of the two islands, then sailed along the east coast of Australia – the first time it had been seen and charted by Europeans. The crew anchored in what is now Sydney, where the botanist Joseph Banks collected many new species of plants and Cook claimed the already inhabited lands for Britain.

Sailing north, the ship collided with the Great Barrier Reef, forcing it to head to a nearby estuary. There, some of the crew came across the Guugu Yimithirr people, who lived in the area, along with an animal the Europeans had never seen before, the *gangurru* (kangaroo).

The long voyage home saw many of the crew die of malaria and dysentery, although they avoided scurvy, as Cook insisted they eat fresh food and nutrient-rich sauerkraut (pickled cabbage).

'We observed much to our surprise that instead of going upon all fours this animal went only upon two legs, making vast bounds.'

Botanist Joseph Banks's account of the crew's encounter with kangaroos

THE FORGOTTEN STORY OF TUPAIA

Tupaia was a brilliant Polynesian navigator who sailed with Cook from Tahiti. His knowledge made him one of the most influential people on board the *Endeavour*. He helped Cook make a map of the Tahitian islands and piloted the ship across the South Pacific. He also introduced the English officers to local Polynesians, who shared their knowledge of edible plants, and when they reached New Zealand, he helped negotiate with local Māori to give the crew food, water and safe harbour.

THE SECOND VOYAGE

In 1772, Cook set out on a second voyage to find the southern continent. The expedition sailed close to the Antarctic coast, but the bitter, sub-zero conditions forced them to return to England in 1775. Although they didn't reach Antarctica, they were the first humans to cross the Antarctic Circle.

THE FINAL VOYAGE

In 1776, Cook's third expedition set out to find the Northwest Passage, in the hope that this might provide a shorter route between the Atlantic and Pacific oceans than the Cape of Good Hope. Unable to find a way through the Arctic ice, Cook decided to sit out the winter in Hawaii. The islanders at first welcomed the captain but relations quickly soured when Cook tried to take a local leader hostage after the theft of one of his ship's boats. During a struggle, Cook was killed on 14 February 1779.

SIBERIA
ALASKA
BRITAIN
Atlantic Ocean
HAWAII
Pacific Ocean
TAHITI
TONGA
AUSTRALIA
NEW ZELAND
CAPE OF GOOD HOPE
Antarctic Circle

First voyage (1768–1771)
Second voyage (1772–1775)
Third voyage (1776–1779)

Bungaree and Matthew Flinders

(C.1775–1830 and 1774–1814)

Captain Matthew Flinders is famed as the brave English sailor who first circumnavigated and mapped Australia, proving that it was a single continent. Less well-known is the Aboriginal guide and navigator, Bungaree, who sailed alongside Flinders. He communicated with people along the way and played an important diplomatic role as they journeyed around the coast.

BUNGAREE AND FLINDERS MEET

Bungaree was born in about 1775 among the Garigal people, five years after Captain Cook had claimed the eastern part of Australia for Britain. Not long after, huge numbers of Aboriginal Australians, who had lived on the continent for tens of thousands of years, were attacked, forced off their land or wiped out by the deadly diseases brought in by the European settlers.

In the 1790s, Bungaree moved to the growing city of Sydney after being driven out of his home along the Hawkesbury River by white settlers. There he became a popular figure among Aboriginal peoples and the European newcomers. By 1798, he had found work on an English ship bound for Norfolk Island, some 1,600 km (1,041 miles) off the coast of Australia. On the voyage, he met the English explorer and navigator Matthew Flinders. Flinders was so impressed by Bungaree that he asked him to join him on his expedition around the unknown coastline of Australia.

THE VOYAGE OF HMS *INVESTIGATOR*

The two men and the rest of the HMS *Investigator* crew departed from Sydney in May 1802. Their aim was to sail around and map the entire coastline of Australia, and to collect plant specimens. Along the way, the voyagers came across a variety of Aboriginal communities, many of whom did not welcome the strange newcomers. Bungaree, however, was a great diplomat and was tasked with negotiating with the local people. Even if he didn't speak their languages, Bungaree found inventive ways to communicate, such as using hand gestures.

Bungaree was also an expert spear fisherman and when the ship's crew went ashore, he helped show them which plants were safe to eat. The expedition took nearly a year and, along the way, many of the crew got sick and died of scurvy or dysentery. The ship also had rotting timbers and extensive leaks, but it finally returned to Sydney in June 1803. It was a remarkable feat of navigation.

AFTER THE VOYAGE

On his return to England, Flinders was captured by the French and imprisoned on the island of Mauritius for six years. While he was there, he managed to send back to England a hand-drawn map of the continent he called 'Australia'. His later book and atlas of maps, titled *A Voyage of Terra Australis*, further helped the name Australia become popular. Because of this, Flinders is credited with having given Australia its name.

Bungaree, meanwhile, remained in Sydney where, dressed in scarlet military uniform, he became a well-known and much-liked figure. He continued to assist on European voyages and, in 1815, was given a large piece of land by the governor of Sydney. Bungaree also helped to support the Aboriginal community and to welcome visitors to Australia and educate them about Aboriginal culture.

REMEMBERING FLINDERS AND BUNGAREE

Throughout Australia, various landmarks, schools, a university and roads are named after Matthew Flinders and there are even statues of his cat, Trim. However, Bungaree's involvement in the exploration of Australia is largely forgotten and, as yet, there are no statues of the great Aboriginal explorer, the first known Australian to circumnavigate the continent on which he was born.

> **'He has always been noted for his kindness of heart, gentleness and other excellent qualities . . . [and] often endangered his life in his efforts to keep the peace within his tribe.'**
>
> Description of Bungaree by Russian explorer Fabian Gottlieb von Bellingshausen

Charles Darwin

(1809–1882)

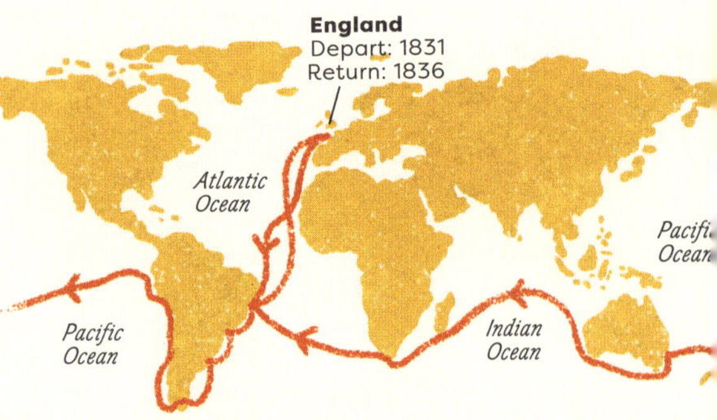

In December 1831, a young Charles Darwin embarked on a fantastic round-the-world voyage on the HMS *Beagle*, collecting plant and animal specimens and observing the wonders of tropical lands. His most famous stop was the Galápagos Islands, where he found plants, birds and tortoises that were uniquely different to anything living elsewhere. His findings led to his ground-breaking theory of evolution and gave us new insights into the amazing diversity of life on our planet.

VOYAGE OF SCIENTIFIC DISCOVERY

The voyage was sponsored by the British government, and its main purpose was to make scientific observations and to survey the coastline of South America. It was planned as a two-year trip, but instead took nearly five years. At first, Darwin suffered terrible seasickness, and conditions were cramped on board the small ship – the naturalist shared a cabin with two other men, and he had to sleep in a hammock that hung over a table.

FOSSILS, TORTOISES AND MARSUPIALS

Fortunately, once the ship reached South America, it made frequent stops and Darwin was able to go ashore to explore and collect specimens. He spent much of his time in the South American wilderness investigating everything from the rainforests of Brazil to fossil-bones in Argentina. After two-and-a-half years, the ship reached the Pacific Ocean and Darwin went on to explore the Andes mountains and the Galápagos Islands, where he made detailed observations about finches, tortoises and mockingbirds. From there, he crossed the Pacific Ocean to New Zealand and Australia (where he marvelled at the marsupials), heading back to England on 2 October 1836.

> **'Here I first saw a tropical forest in all its sublime grandeur . . . I never experienced such intense delight.'**
>
> Darwin explores Brazilian rainforests for first time, February 1832

DIVERSITY OF LIFE

By the end of the voyage, Darwin had filled several notebooks with his observations of plants and animals, and collected over 1,500 specimens. He eventually went on to publish a very important book called *On the Origin of Species*. In the book, he outlined his revolutionary theory that the billions of species of plants and animals on our planet evolved (changed or developed) over millions of years through 'natural selection' – a process that is still at work today.

ALEXANDER VON HUMBOLDT

The young Charles Darwin was inspired and guided by the work of the German naturalist Alexander von Humboldt. Both explored the plants and animals of South America and were astonished by its diversity. Von Humboldt travelled throughout the Americas between 1799 and 1804, mapped the distribution of plants on three continents and tracked what became known as the Humboldt Current in the Pacific Ocean.

SLAVERY AND SCIENCE

Now largely lost to history is the role of the slave trade in European voyages of scientific discovery in the 18th century. Specimens were often carried on slave ships, the work of European naturalists was sometimes funded by slavery, and it was not uncommon for scientists to purchase an enslaved assistant.

GRAMAN KWASI

One former enslaved person who unusually was recognised for his work was Graman Kwasimukambe (also known as Kwasi), a Ghanaian-born enslaved person who had been transported from West Africa to Suriname, a Dutch colony in South America. He was known for his *obeah* (skill in medicinal plants and spiritual knowledge), which he used to heal enslaved people and to treat Europeans for money. In 1730, Kwasi discovered a plant that could treat fever and keep parasites away. In 1762, Swedish botanist Carl Linnaeus, who devised the system by which we name all plants and animals today, named the plant *Quassia amara* in his honour and it continues to be used in herbal teas and medicines today.

Ocean Record-Breakers

Sailors still embark on great oceanic voyages, just as pioneering seafarers did centuries before. These days, however, circumnavigating the globe can take a matter of days rather than years. Modern yachts are streamlined and fast, and women have proven that they are just as likely as men to smash world records.

SOLO SAILING

During the Age of Discovery, European sea captains often crossed oceans in large ships carrying many crewmembers. Today, great ocean liners can carry thousands of people, although some brave sailors race across the vast oceans in much smaller vessels, entirely on their own. The first person to sail around the world alone was Canadian seaman and adventurer Joshua Slocum. Setting sail in 1895 from Massachusetts, USA, his extraordinary journey took three years, and he faced ferocious storms, treacherous reefs, encounters with pirates, attacks by local tribes and shipwreck.

In 1967, British sailor Sir Francis Chichester became the first person to sail solo around the world making just one stop. The first person to sail solo around the world nonstop was Sir Robin Knox-Johnston, also from Britain. His historic voyage took more than ten months between June 1968 and April 1969.

ALL AROUND THE WORLD

In 1978, Polish sailor Krystyna Chojnowska-Liskiewicz became the first woman to sail single-handedly around the world. An experienced sailor and ship engineer, Chojnowska-Liskiewicz wanted to prove that a woman was capable of sailing across the globe alone. Setting off from the Canary Islands, the Polish mariner faced many dangers, including navigating her custom-made yacht over the razor-sharp corals of the Great Barrier Reef and almost colliding with other boats. She nonetheless completed her record-breaking voyage in 401 days.

SPEED RECORD

British sailor Ellen MacArthur had been sailing boats since the age of four. By the time she was in her twenties, she was determined to beat the record for a nonstop solo voyage around the world. In November 2004, she set off from the official starting point of France. She made fast progress in her racing yacht, although she badly burnt her arm fixing a generator on the boat. Despite not being able to sleep for more than 20 minutes at a time, MacArthur eventually made it back to France in 71 days, breaking the previous record set by French sailor Francis Joyon by 1 day and 8 hours. Francis Joyon eventually beat MacArthur's record, as did fellow French sailor François Gabart in 2017, sailing solo around the world in a racing yacht in 42 days and 16 hours.

'I knew then that I wanted to sail around the world. As a kid, that was the goal. I had no idea how to get there . . . but I knew that was what I wanted to do at some stage.'

Ellen MacArthur

Joshua Slocum's boat the Spray sailing through stormy seas

TEEN ADVENTURERS

Teenage sailors have also taken up the challenge of sailing around the world. In 2009, 17-year-old Zac Sunderland from California became the first person under the age of 18 to sail solo around the globe. His record, however, was short-lived, as six weeks later, an even younger 17-year-old from the UK, Michael Perham, achieved the same feat. Two years before, Perham had successfully sailed across the Atlantic Ocean at just 14 years old. If this wasn't impressive enough, in 2010, 14-year-old Laura Dekker from Holland began a solo circumnavigation of the world, which she completed at the age of 16, the youngest person ever to do so.

Zac Sunderland, Laura Dekker and Michael Perham

The Ocean Deep

Oceans cover more than 70 per cent of the Earth's surface, but their depths are still a mystery. Humans have built contraptions to take them underwater, but few have ventured to deepest parts of the sea. Scientists have only recently discovered that the ocean floor is made up of vast mountain ranges, deep ridges and valleys, and an entire ecosystem of strange creatures that can survive without sunlight.

DIVING BELLS

Humans have always swum in the sea, diving beneath the surface to fish or to collect sponges, valuable shells, coral, or even to salvage goods from shipwrecks. The earliest type of equipment to transport people to the sea depths were diving bells – hollow objects with an open bottom. They were first used as early as the 4th century BCE. Centuries later, in 1691, the English astronomer Edmond Halley designed a diving bell that could remain submerged for up to 90 minutes.

DIVING SUITS

In the early 1700s, the English inventor John Lethbridge built one of the first known diving suits – an air-filled, pressure-proof barrel with a glass viewing hole. It was used to salvage valuable cargo from sunken ships, including several tonnes of silver.

SUBMERSIBLES

In 1620, the Dutch inventor Cornelis Drebbel designed and constructed what is thought to be the first submarine. It was built of leather and wood, and made a number of trips under the surface of the River Thames, England, reaching a depth of 4.6 m (15 ft).

Drebbel's submarine was propelled by oarsmen and air was supplied by two tubes with floats to keep one end above water.

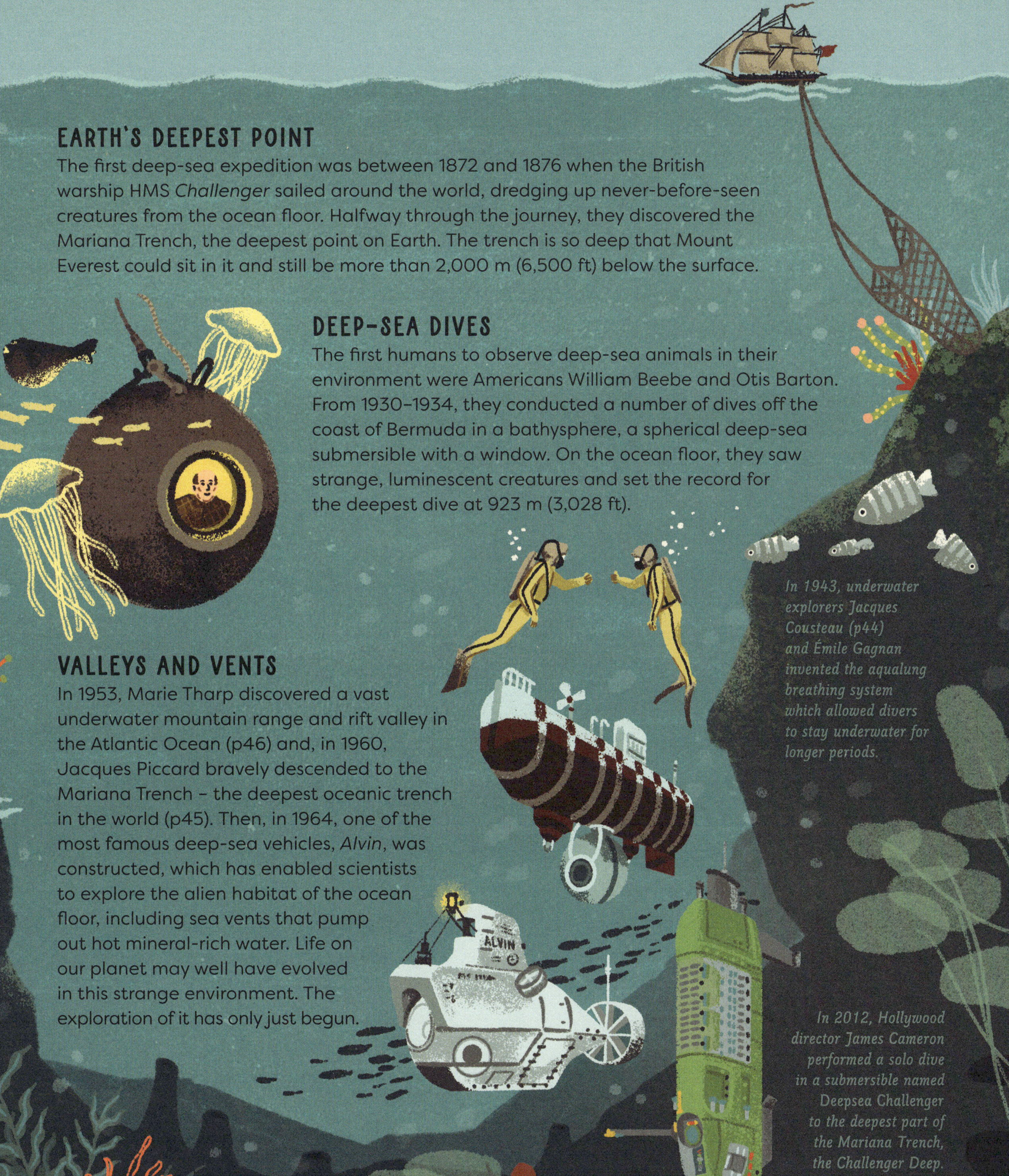

EARTH'S DEEPEST POINT

The first deep-sea expedition was between 1872 and 1876 when the British warship HMS *Challenger* sailed around the world, dredging up never-before-seen creatures from the ocean floor. Halfway through the journey, they discovered the Mariana Trench, the deepest point on Earth. The trench is so deep that Mount Everest could sit in it and still be more than 2,000 m (6,500 ft) below the surface.

DEEP-SEA DIVES

The first humans to observe deep-sea animals in their environment were Americans William Beebe and Otis Barton. From 1930–1934, they conducted a number of dives off the coast of Bermuda in a bathysphere, a spherical deep-sea submersible with a window. On the ocean floor, they saw strange, luminescent creatures and set the record for the deepest dive at 923 m (3,028 ft).

In 1943, underwater explorers Jacques Cousteau (p44) and Émile Gagnan invented the aqualung breathing system which allowed divers to stay underwater for longer periods.

VALLEYS AND VENTS

In 1953, Marie Tharp discovered a vast underwater mountain range and rift valley in the Atlantic Ocean (p46) and, in 1960, Jacques Piccard bravely descended to the Mariana Trench – the deepest oceanic trench in the world (p45). Then, in 1964, one of the most famous deep-sea vehicles, *Alvin*, was constructed, which has enabled scientists to explore the alien habitat of the ocean floor, including sea vents that pump out hot mineral-rich water. Life on our planet may well have evolved in this strange environment. The exploration of it has only just begun.

In 2012, Hollywood director James Cameron performed a solo dive in a submersible named Deepsea Challenger to the deepest part of the Mariana Trench, the Challenger Deep.

Jacques Cousteau

(1910–1997)

Jacques Cousteau was a world-famous French scientist, filmmaker, photographer and explorer of the sea. He pioneered a host of diving inventions, including a device that helped divers breathe underwater, and he sailed the world, exploring beneath the waves. He was also a passionate campaigner for the preservation of ocean habitats, and educated millions about the Earth's oceans and its great mysteries.

YOUNG DREAMS

As a child, Jacques Cousteau dreamed of becoming a pilot. However, when he was 13, he broke both his arms in a car-crash. To build his strength after the accident, he went swimming in the Mediterranean Sea. Having borrowed some goggles from a friend, he was so amazed by what he saw underwater, he knew he wanted to explore the sea further.

FILL YOUR LUNGS

In 1943, he met the French engineer Émile Gagnan, and together they experimented with devices that would help divers spend longer in the water and move more freely. One of their inventions included a breathing apparatus called an aqualung, which feeds air to the diver from oxygen tanks. It allowed Cousteau and Gagnan to explore and film underwater.

AMAZING UNDERSEA WORLDS

In 1950, Cousteau was given a ship called *Calypso* from which he explored the oceans and carried out research. People around the world read his books and watched his undersea adventures in films and in his famous television series *The Undersea World of Jacques Cousteau*. In the 1970s, he explored the Southern Ocean – the last untouched ocean of the world – and later turned his attention to the environment and what could be done to better protect the seas.

> 'From birth, man carried the weight of gravity on his shoulders. He is bolted on Earth. But man has only to sink beneath the surface and he is free.'
>
> Jacques Cousteau

Jacques Piccard

(1922–2008)

›››››››››››››››››››››››››››

The Swiss engineer and oceanographer Jacques Piccard is best known for his record-breaking exploration of the deepest known part of the planet, the Mariana Trench. It was a remarkable feat requiring expertise, determination and immense courage to descend to the very bottom of the ocean, never before reached by humans.

THE BATHYSCAPE

Jacques' father, August Piccard, was also an engineer and adventurer, and twice beat the record for flying higher than anyone else in a hot-air balloon. He later turned his attention to the ocean depths and Jacques helped his father develop a machine called a bathyscape which helped people dive deep underwater.

ROCK BOTTOM

Jacques and his father went on to build and test-dive three bathyscapes which reached the record depth of 6,000 m (19,685 ft). In 1958, the US Navy decided to fund Piccard's research as they were interested in underwater research and were looking at ways submarines could be rescued. Using a bathyscape they named *Trieste*, Jacques and his team planned to descend to the deepest part of the ocean.

'The bottom appeared light and clear, a waste of snuff-coloured ooze.'

Jacques Piccard's description of the Mariana Trench

LIFE IN THE OCEAN DEEP

On 23 January 1960, Jacques and Lieutenant Don Walsh of the US Navy made history when they descended 10,911 m (35,797 ft) to the floor of the Mariana Trench in the Pacific Ocean. After four hours, the crew suddenly heard a loud noise as one of their Perspex windows cracked. Showing incredible bravery, they continued their descent, reaching the bottom 40 minutes later. They were amazed to see what appeared to be flatfish floating by their porthole, despite biologists claiming that no fish could live at such depths.

The water pressure at the bottom of the trench is around 1,000 times greater than it is at sea level – so great that your bones would dissolve.

Marie Tharp

(1920–2006)

Until the 1950s, most people thought the bottom of the ocean was flat and featureless. But in 1953, the young geologist Marie Tharp produced one of the first maps of the ocean floor, revealing mountains and a huge valley in the middle of a gigantic ridge running across the Atlantic Ocean. This was a revolutionary discovery – accomplished at a time when few women were able to work in the sciences – and provided important evidence about the development of the planet.

A CAREER IN SCIENCE

In 1942, American graduate Marie Tharp was given the opportunity to study for a master's degree in geology. This was during the Second World War, when women were able to pursue careers that were usually dominated by men because so many men were serving in the armed forces. She and student Bruce Heezen went on to became part of a research project to map the ocean floor.

A research vessel was sent to measure the depths of the ocean. However women were not allowed to sail on research boats. Instead, Marie collected the data and meticulously plotted out the measurements using only pens and rulers.

'It was a once-in-the-history-of-the-world opportunity . . . especially for a woman in the 1940s.'

Marie Tharp

SEISMIC DISCOVERY

Over the next 20 years, Marie produced detailed maps of the oceans, including, in 1977, a map of all the world's oceans. The maps revealed that the ocean floor was covered in mountains, trenches and canyons. Marie also identified a deep rift valley running along the Mid-Atlantic Ridge, which, scientists realised, was also the epicentre of many earthquakes.

This supported the theory of plate tectonics – the idea that continents move over time and the oceanic crust is spreading apart. Marie's male colleagues initially dismissed her findings as 'girl talk' and refused to believe it for a year. The discovery of the rift valley – now considered Earth's largest feature – was nothing short of seismic, although at the time, Marie Tharp received little or no credit.

Dr Cindy Lee Van Dover

(1954–)

Dr Cindy Lee Van Dover was the first and only female pilot of the research submersible *Alvin*. Built in 1964, the US Navy-owned vessel enables scientists to explore what lies in the deepest parts of the ocean. Cindy has led many *Alvin* expeditions to study this mysterious environment, where life on our planet may well have begun.

FIRST DIVES

Ever since she was a child, Cindy Lee Van Dover longed to explore the deep sea. In 1982, she embarked on the first expedition to study hydrothermal vents in the East Pacific Rise – openings in the sea floor which spray out hot, mineral-rich water. She then had her first dive in the deep-sea submersible *Alvin* in 1985, and in 1990 became its first female pilot.

STRANGE SPECIES

Alvin can carry a pilot and two passengers and is equipped with two robotic arms. It is built to withstand the crushing pressures of the deep sea and has portholes and lights so scientists can see the incredible creatures that live in the murky waters. Around 300 animal species have been discovered – from eel-like fish and furry snails to sea vent tubeworms. *Alvin* has also been used to locate bombs on the seabed and has made several dives to the shipwreck of the *Titanic*.

DEEP-SEA VENTS

Cindy has made more than 200 dives to the ocean floor, collecting specimens and making important discoveries about deep-sea vents and how organisms thrive in this extreme environment.

Deep-sea vents have been around since the oceans first formed, long before life began on Earth and life itself may well have evolved in this deep-sea environment.

Overland EXPLORERS

Humans have ventured across every continent on the planet. They have trekked through swamps, ridden camels across deserts and sped along railway tracks. Throughout our history, people have explored new lands to find food, to conquer people and places, trade goods or escape danger. Some explorers travelled long distances overland for religious reasons or out of scientific curiosity, while others sought adventure, riches or simply to know what lay beyond the next bend.

Marco Polo

(1254–1324)

At the age of 17, Marco Polo left his home in Venice and travelled to China and the magnificent court of the Mongol emperor Kublai Khan. He saw things no other European had ever seen before and returned to Italy 24 years later. His hugely successful book about his extraordinary adventures inspired many others to explore the world.

ADVENTURES ON THE SILK ROAD

Marco Polo's father and uncle had made their fortune trading silks, spices and jewels. They had travelled along the Silk Road (p10) to China, where they established friendly relations with Kublai Khan, the ruler of the world's then biggest empire. In 1271, they decided to head to China again, this time taking Marco with them.

VAST MARBLE PALACE

The journey was treacherous as they ventured overland across parched deserts and icy mountains, their jewels hidden within the seams of their coats for safety. Along the way, they visited many great cities. Three years later, they finally reached Shangdu, the summer palace of Kublai Khan, where Marco marvelled at the amazing marble buildings and beautiful gardens. The Mongol emperor was impressed with the young Marco and soon employed him, sending him on fact-finding missions across China, Myanmar (Burma) and India.

SHARP-CLAWED SERPENTS

After 16 years in China, the Polos set out to return to Venice, leaving China with a fleet of 14 ships. They encountered terrible weather, and monsoon winds forced them to spend five months on the island of Sumatra, now in Indonesia. Marco later wrote of the strange beasts he came across, including elephants, monkeys and crocodiles, which he described as giant, sharp-clawed 'serpents' that could 'swallow a man . . . at one time'. On seeing some rhinoceroses, he described them as enormous unicorns.

BESTSELLING BOOK

On returning to Venice, Marco was taken prisoner during a war with Venice's rival Genoa. While in captivity, he related all the amazing sights he had seen to the writer Rustichello, who wrote them down in a book called *The Travels of Marco Polo*. He described the Mongol Empire's great cities, its paper money (which was unknown in the West) and postal system, although he did exaggerate about some things he saw, leading some to doubt his account. Nevertheless, his book about strange new lands in the East excited people and encouraged other Western explorers to travel, including Christopher Columbus, who took a copy of the book on his voyages a century later.

Some of Marco Polo's tales were greatly exaggerated, like his description of the mythical Gryphon bird.

> 'It was for all the world like an eagle, but one indeed of enormous size . . . And it is so strong that it will seize an elephant in its talons and carry him high into the air and drop him so that he is smashed to pieces.'
>
> Tall tales of Marco Polo

RABBAN BAR SAUMA (1220–1294)

Born in Beijing, the Christian monk Rabban bar Sauma made a similar epic journey in 1275 to 1288, but he travelled in the opposite direction, from China to Europe. His name may be less well-known, but his achievements are no less extraordinary.

Sometime around 1275–1280, Sauma left China to make a pilgrimage to Jerusalem. When visiting Baghdad, he was sent by a Mongol ruler on a mission to meet with monarchs in western Europe in the hope they would help the Mongols expel Muslims from Jerusalem. Sauma made his way to Constantinople and was dazzled by the city's beauty. From there, he sailed to Sicily, where he watched the volcano Mount Etna erupt, then travelled to France, where he met the French king Philip the Fair and the English king Edward I, who treated him to a lavish feast. He then went to Rome, where he met the pope. He didn't make any formal alliances, but his travels laid the groundwork for further contact and trade between Asia and the West.

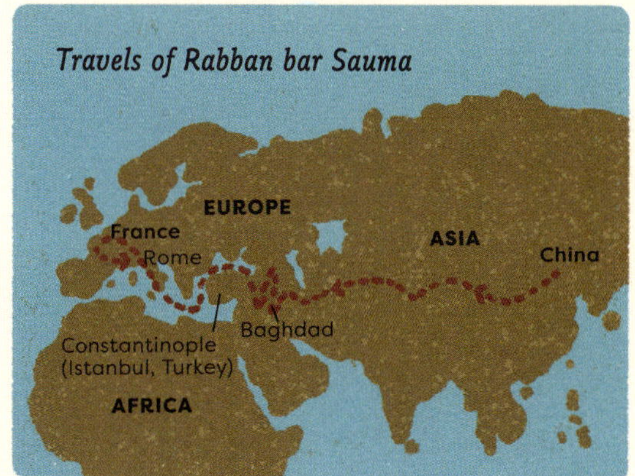

Ibn Battuta

(1304–1369)

Ibn Battuta, a Moroccan scholar and explorer, was one of the greatest travellers in history. Over the course of 29 years, he travelled some 117,000 km (73,000 miles) on an odyssey across the Islamic Empire as far as China – further than any other explorer in the period. Passing through more than 40 countries, he faced shipwrecks, pirates, bandits and tyrannical rulers.

A MUSLIM PILGRIMAGE

In June 1325, the 21-year-old Ibn Battuta set out from Tangier, Morocco, intending to complete his Hajj – a Muslim pilgrimage to the holy city of Mecca – and to learn Islamic law. He began his journey alone, riding on a donkey, but soon joined a caravan of camel-riding travellers as he crossed North Africa. Despite having company, he felt lonely and homesick, and on reaching Algeria, he caught such a severe fever that he had to tie himself to his saddle to stop from falling to the ground. During his journey to Cairo in Egypt, he married twice, and he went on to wed and divorce ten times during his travels!

Mecca is revered in Islam as the birthplace of the Islamic prophet Muhammad.

A CITY OF BEAUTY

From Cairo, Battuta travelled to Jerusalem, where he was dazzled by the Dome of the Rock (an important Islamic shrine) and then by the Umayyad Mosque in Damascus, Syria, which he described as 'the city that surpasses all other cities in beauty'. In 1326, a year and a half after setting out on his adventure, he finally reached Mecca and completed his pilgrimage.

> 'I set out alone . . . swayed by an overmastering impulse . . . to visit these illustrious sanctuaries.'
>
> Ibn Battuta

WANDERLUST

The journey had been a long and exhausting one but, along the way, Battuta had discovered a passion for travel. He loved learning about new countries and new peoples, so, instead of returning home, he continued his adventures for almost 30 years! From Mecca, he travelled north to Baghdad in modern-day Iraq, where he delighted in the city's public baths. Later, he trekked across Yemen and sailed south along the east African coast, dodging pirates along the way.

CAVE DWELLER

In 1332, Battuta arrived in Constantinople in modern-day Turkey, where he met the Byzantine emperor and was paraded around the city on horseback. He then headed to India and worked as a judge for the Sultan of Delhi for a few years. However, the sultan, who was notoriously cruel to his enemies, accused Battuta of disloyalty and he was forced to live in a cave for five months until, much to his relief, he was sent to the Mongol court in China in 1341.

DANGER LURKS

Battuta's journey to China was fraught with danger. Hindu rebels kidnapped him and robbed him of everything except his trousers. Then, his ships were caught in a violent storm in the Indian port of Calicut, and many in his party were killed. Fearing for his life, he crossed the Indian Ocean to the islands of the Maldives, eventually arriving in China in 1345.

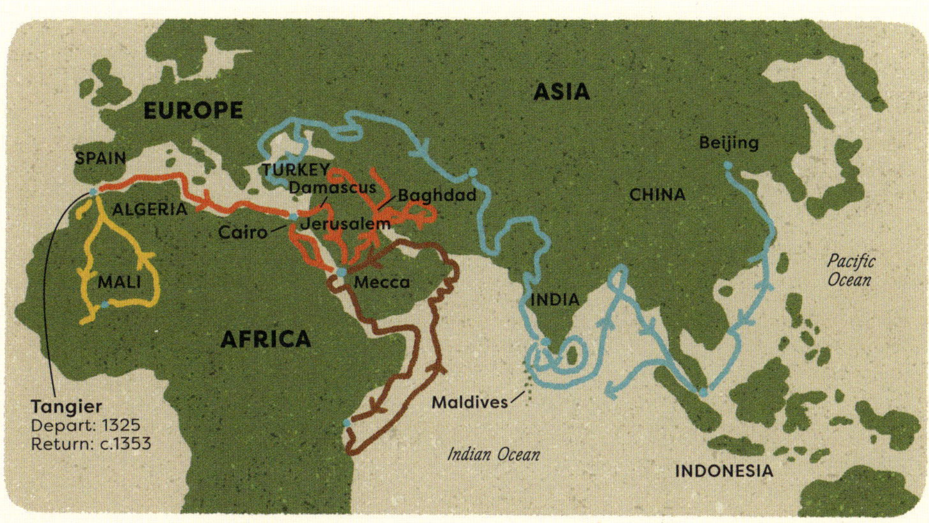

The travels of Ibn Battuta — 1325–27 — 1330–32 — 1332–46 — 1349–53

TRAVEL BOOK

On his return voyage, Battuta travelled, among other places, to Indonesia, Moorish Spain and across the Sahara Desert to the Mali Empire. He finally returned to Morocco for good in around 1353. Once home, the sultan ordered him to compile a book of his adventures. He recounted his tales to a writer who produced the now-famous travelogue known as the *Rihla* (Travels).

CHRISTIAN PILGRIMAGES

Just as Muslims conducted Hajj to the sacred city of Mecca, Christians also went on pilgrimages to sacred sites such as the Holy Land, which includes Jerusalem, Nazareth, Bethlehem and other places linked to Jesus and his apostles. Pilgrimages were the most common reason for long journeys in the Medieval period.

The Spanish conquistadors (conquerors) were some of the first Europeans to explore Central and South America. Carrying guns and fighting on horseback, they swiftly overpowered the indigenous peoples, looted their treasure and destroyed their civilisations, including the mighty empires of the Aztecs and Incas.

Conquistadors

BUILDING EMPIRES

After Columbus sailed to America in 1492, Spain saw an opportunity to expand its wealth and power by adding new territories to its empire, finding new trade routes to the East and converting people to Christianity. Many of those who sailed to the Americas also longed for adventure and were lured by tales of gold, silver and other treasures.

HERNÁN CORTÉS AND THE AZTECS

One of the most infamous conquistadors was Hernán Cortés (1485–1547). In 1519, he took a fleet of ships carrying 500 armed soldiers to the east coast of Mexico, before marching to Tenochtitlan, the capital of the region's dominant civilisation, the Aztecs. The Aztec emperor Moctezuma welcomed them at first. However, Cortés later put Moctezuma under arrest and slaughtered hundreds of Aztec nobles. Moctezuma too was eventually killed.

Meanwhile, the European disease of smallpox, which the Spaniards had brought with them (and to which the Spanish were mostly immune), ripped through the city, killing a quarter of the population. In 1521, Cortés attacked Tenochtitlan, causing another 100,000 deaths. After months of bombardment, the once-great city lay in ruins. This led to the collapse of the Aztec Empire, and Cortés became the ruler of what was called 'New Spain'.

FOUNTAIN OF YOUTH

The conquistador Juan Ponce de León (1474–1521) was also encouraged by the Spanish crown to discover new lands. In 1513, he set sail from Puerto Rico and landed on the coast of what is now Florida. Legend has it that he looked for the island of Bimini – home to a magical spring that provided eternal youth for anyone who bathed or drank from it.

GOLD AND SILVER

Another Spanish conquistador enticed by the promise of adventure and riches was Francisco Pizarro (1478–1541). In November 1532, he and 168 armoured soldiers on horseback entered the Peruvian city of Cajamarca, where they captured Atahualpa, the leader of Inca Empire, which was then the largest in the world, covering a huge swathe of South America. Pizarro promised to spare Atahualpa's life if he paid a ransom. The Inca leader dutifully filled a large room with gold and silver, but the Spaniards broke their promise and killed him. Atahualpa's death led to the end of the Inca Empire. Other expeditions extended Spanish rule into Ecuador, Colombia and Chile.

CITY OF GOLD

The Spanish continued to explore the Americas for the rest of the 16th century, some in search of 'El Dorado' – a mythical city of gold and precious jewels. They never found the city, but they did discover silver mines in Mexico and Bolivia and sent ships laden with silver back to Spain, making it the richest country in Europe.

LEGACY OF THE CONQUEST

The European colonisation of the Americas was to have far-reaching consequences. Foods brought from and to the New and Old Worlds brought economic benefits and transformed the diets of millions of people across the globe, with chillies, tomatoes, maize, cocoa, potatoes, sweet potatoes and sunflowers going east; wheat, grapes, olives, sugar cane, rice, coffee, sheep, cattle, chickens, pigs and horses going west.

However, it also had a devastating effect on the local people and cultures. Settlers brought with them European diseases, which killed as much as 80 per cent of the native population over the course of the 16th century. Slaves from Africa were increasingly shipped to the Americas to replace the millions who had died, and the great civilisations and temples of the indigenous peoples were destroyed.

Estevanico

(C.1500–1539)

Estevanico was the first person from Africa to explore North America. Captured and sold into slavery, he was taken on a gruelling expedition to explore lands between Florida and Mexico. After facing violent hurricanes, fights with local people and near-starvation, Estevanico was one of just four survivors of the ill-fated mission. They continued to wander southwestern North America for eight years, before finally reaching Mexico in 1536.

CONQUISTADORS

Estevanico – also known as Esteban de Dorantes or Mustafa Azemmouri – was born in around 1500 in the Portuguese-controlled town of Azemmour, Morocco. Here, he was enslaved and sold to a Spanish nobleman who joined an expedition led by the Spanish conquistador Pánfilo de Narváez to conquer lands west of Florida in North America.

OVERLAND TREK

After a difficult journey on five overcrowded ships from Spain to Hispaniola, the expedition moved on to Cuba. In February 1528, the ships and about 400 men set off for Mexico, but hurricanes forced them to stop near Tampa Bay in Florida. After claiming the territory for Spain, Narváez took 300 men and 40 horses to explore the land, planning to meet up with the remaining fleet later. After three terrible months, during which at least 40 men perished as they trekked through swamps and fought with local Native Americans, they reached Apalachee Bay, Florida.

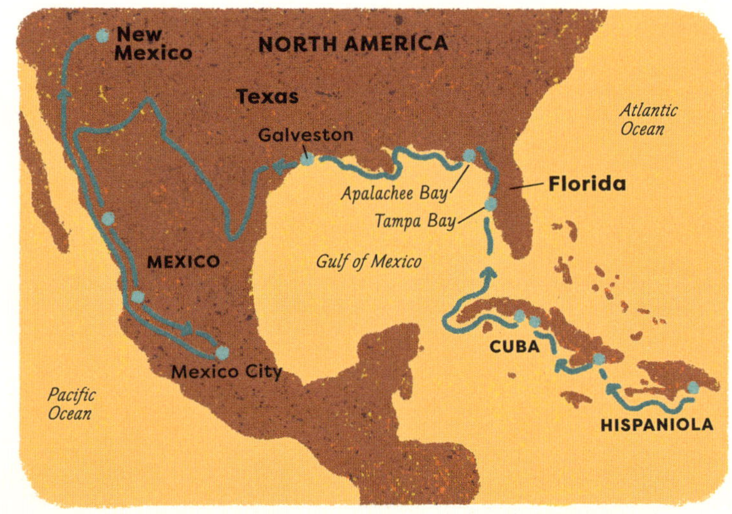

Estevanico's route

NARVÁEZ VANISHES

With no sign of the ships, they decided to build five barges out of whatever they could find, melting down metals from bridles and stirrups, making ropes from horsehair and sewing shirts together for sails. In late September, they sailed along the coast towards Mexico. Strong currents, however, separated the vessels and Narváez's barge was pushed out into open sea, never to be seen again.

Estevanico's vessel capsized, but its occupants, along with another barge, made it to Galveston Island in Texas. Most of these survivors died during the long, cold winter leaving just 15 men from the expedition still alive by the following spring. The remaining men made it to the mainland, where they were captured by local inhabitants. Over the following months, more of the group died, leaving just four men, including Estevanico. In 1534, they escaped and made their way inland into Texas and northern Mexico – the first non-natives to enter the American West.

> 'It was in November, bitterly cold, and we in such a state that every bone could easily be counted, and we looked like death itself.'
>
> Cabeza de Vaca, one of the four survivors, describing their arrival on Galveston Island

MAGICAL POWERS

As they travelled, they posed as medicine men. People were struck by their unusual appearance and believed they had magical powers and could cure people of illness. They were showered with food and gifts, and Estevanico – who was an excellent linguist – communicated with local tribes, either by learning their languages or by using sign language. This helped them to travel safely and find the best routes. Two years later, after a journey of more than 3,000 km (1,800 miles) over deserts and mountains, they finally reached the Spanish settlement at Sinaloa on the west coast of Mexico, before travelling on to Mexico City.

SEVEN GOLDEN CITIES

Estevanico was then sold as an enslaved person to the Viceroy of New Spain and sent on an expedition to the fabled Seven Golden Cities of Cibola, where it was rumoured that the streets were paved with gold. Estevanico travelled ahead and on entering a village in New Mexico, was killed by the inhabitants. No one saw how he died – he perhaps mistakenly scared the villagers, or some like to think he faked his death so he could at last secure his freedom.

Map of the fabled cities of Cibola

Francisco de Orellana

(1511–1546)

In 1541, Spanish conquistador Francisco de Orellana became the first European to sail along the entire length of the Amazon River as part of an expedition to search for spices and gold. A huge number of people and animals slogged through dense jungles and swamps, with thousands dying of disease and starvation along the way.

The term El Dorado is Spanish for 'the golden one'. Rumours of this legendary lost city go back hundreds of years and many European explorers travelled to the New World in search of it.

YOUNG ADVENTURER

As a teenager, Orellana travelled from Spain to South America, where he helped his cousin Francisco Pizarro overthrow the Inca Empire in Peru. During the fighting, he lost one eye, and thereafter wore an eye-patch. In March 1541, Orellana joined an expedition, led by Pizarro's brother Gonzalo Pizarro, to search for the much-prized spice of cinnamon and El Dorado, a mythical land said to be abundant with gold.

TANGLED SWAMPS

Setting off from Quito in northern Ecuador, the party included 340 soldiers, about 4,000 enslaved Indigenous South Americans and a variety of animals, including horses and hunting dogs. They soon encountered tangled swamps, dense jungles, insects and poisonous snakes. Supplies were also running dangerously low and many of the slaves were struck down by smallpox.

On reaching the River Coca, Orellana was sent ahead of the main party to look for food, sailing on a makeshift sailing vessel with around 50 soldiers. By this time, the travellers were so hungry that they were forced to boil the leather from their saddles for a Christmas meal. Instead of returning to Pizarro, the party agreed it would be safer to continue downriver.

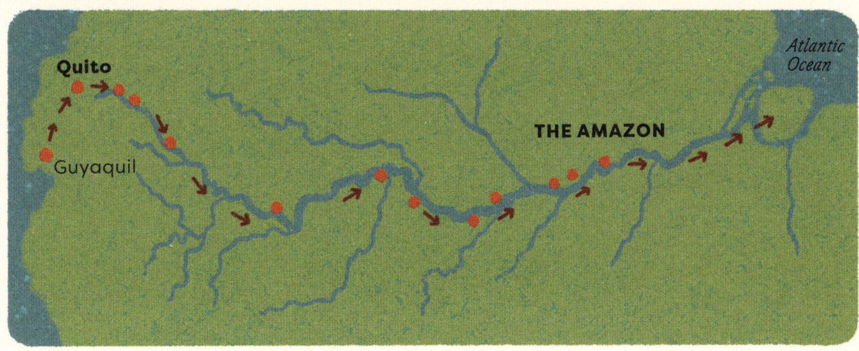

Orellana's route along the Amazon

HALLUCINATIONS

For the first few days, Orellana's group passed through an empty landscape, so close to death they began to hallucinate. They finally came across some Indigenous people who saved their lives and gave them food. Over the coming months, the travellers would meet many other tribes, some of whom were friendly whilst others attacked the strange-looking Spaniards. Drifting with the strong current, the party reached the Napo River on 3 June 1542 and finally came to a gigantic confluence of two rivers, marking the beginning of the Amazon River.

> 'We chose what seemed to us the lesser of two evils. Trusting to God we would go on and follow the river and either die or see what marvels lay ahead.'
>
> Francisco de Orellana

MIGHTY AMAZONS

On 24 June 1542, the group were attacked by a fierce band of women, probably from the Tapuya tribe. According to the priest who wrote a book about the expedition, the women fought with bows and arrows and 'did as much fighting as ten Indian men'. The Spaniards likened these women to the Amazons, who were female warriors of Greek mythology. The story of these women became famous, which led to the river and its region being named in honour of them.

RETURN EXPEDITION

The party eventually made it to the Atlantic Ocean on 26 August 1542, before sailing to Cubagua Island off Venezuela and returning to Spain in 1543. Two years later, Orellana set out to conquer the Amazon, sailing with four ships and hundreds of settlers. While scouting for a place to live, some of his men were attacked by locals and Orellana died of illness shortly after.

Meriwether Lewis, William Clark and Sacagawea

(1774–1809, 1770–1838 and c.1788–1812 or 1884)

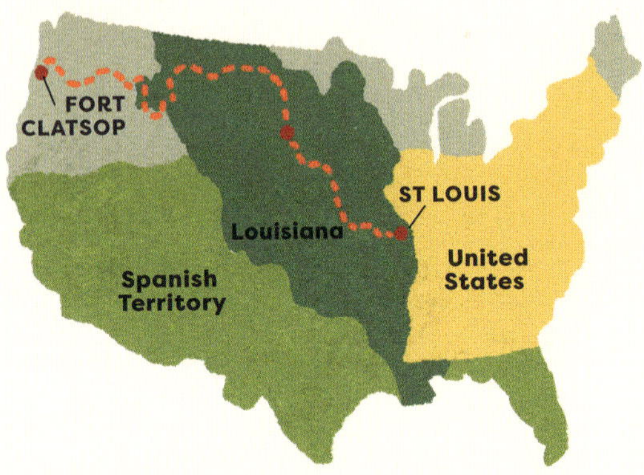

Lewis and Clark expedition (1804–1806)

In 1804–1806, Meriwether Lewis and William Clark led a historic journey of discovery across the wilderness of North America from St Louis in Missouri to the Pacific Coast. They survived illness, snakebites and attacks by grizzly bears. Along the way, they enlisted the help of a young Shoshone Native American woman, Sacagawea, who shared her knowledge and helped establish friendly relations with the many local tribes.

LOUISIANA PURCHASE

In 1803, in a deal known as the Louisiana Purchase, the US president Thomas Jefferson acquired a vast amount of land west of the Mississippi River that almost doubled the size of the United States. Shortly after, the president planned an expedition to explore the new territory as far as the west coast and to gather knowledge about the Native Americans living along the route.

EXPEDITION OF DISCOVERY

To lead the expedition, the president chose his personal secretary, Lewis Meriwether, who was an officer in the US Army and spoke several Native American languages, and William Clark, a retired army officer and friend of Lewis. In May 1804, Lewis and Clark and a team of about 40, including Lewis's Newfoundland dog Seaman, set off from St Louis, Missouri, in three boats. They moved slowly up the river, travelling westward, reaching North Dakota in early November. After narrowly avoiding a fight with the Sioux chief Black Buffalo and his band, the expedition built a camp and spent the winter living amicably alongside the Mandan tribe.

SACAGAWEA

During the winter of 1804, they met Sacagawea, a Shoshone Native American woman and her fur-trader husband Toussaint Charbonneau. They joined the expedition and helped Lewis and Clark to communicate with local tribes, secure essential supplies and horses, and identify plants and herbs that they could eat. As they travelled, Sacagawea carried her baby son on her back. From April 1805, the party continued travelling west on foot and by canoe, traversing mountains and rivers with very little food, finally reaching the west coast in late November, where they built a small fort, Fort Clatsop.

> **'They collect the wild fruits and roots, attend to the horses . . . cook, dress the skins and make all the apparel, collect wood and make their fires, arrange and form their [teepees], and when they travel, pack the horses and take charge of all the baggage.'**
>
> Lewis and Clark on Sacagawea and the role of a Shoshone woman

BEAR ATTACK

During their trailblazing journey they encountered several dangerous animals. Over the course of 24 hours, Lewis was almost bitten by a rattlesnake, attacked by a wolverine and charged by a bison. In May 1805, they came across a grizzly bear, which Clark described as a 'very large and a terrible looking animal', and in June 1805, when Lewis was out hunting, a bear rushed towards him. With no time to reload his rifle, his only option was to jump in a nearby river. Luckily for Lewis, the bear ran away!

Lewis and Clark were amazed to discover many animals for the first time, such as the grizzly bear, coyote, prairie dog and jackrabbit.

AMERICAN WEST

After wintering in Oregon, the party made their way back east, arriving in St Louis in September 1806, more than two years after they had first set out. They returned with detailed maps of the mountains, rivers and lands they had seen, along with journals documenting more than 200 new species of animals and plants, many of which were previously unknown to Americans. The journey also provided information about the cultural practices and languages of Native American tribes, and marked the first step in American settlement of the West and the dispossession of the Native Americans.

(C.1798–C.1866)

Former enslaved person James Beckwourth rose to become a mountain man and explorer and one of the most influential figures in the American West. In the 1820s, he joined a gruelling fur-trapping expedition in the Rocky Mountains, then lived amongst the Crow people for six years, and later discovered a safe route through the Sierra Nevada mountains to the gold-rich regions of California.

SLAVERY

Born in Virginia in around 1798, Beckwourth was the child of a white plantation owner and a black, probably enslaved, woman. After moving to St Louis, Missouri, his father freed him from slavery and set him to work as an apprentice to a blacksmith.

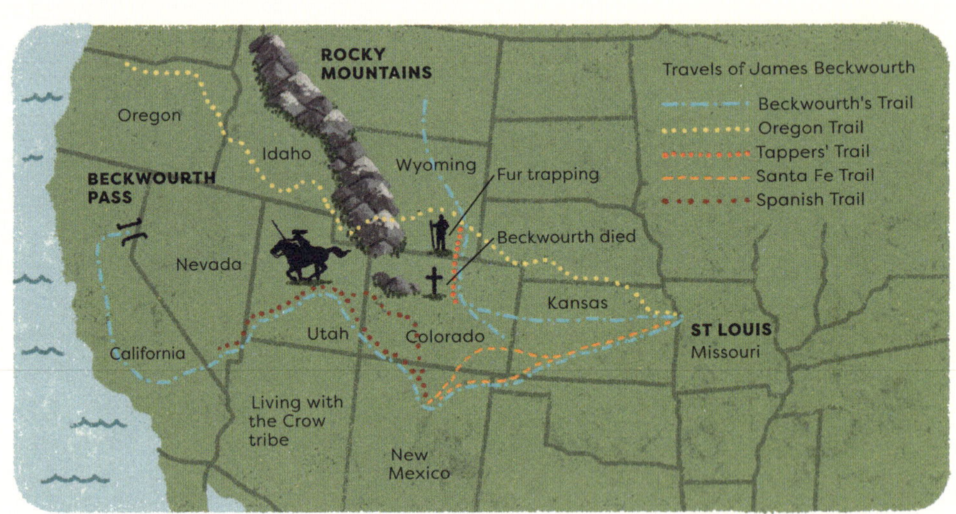

MOUNTAIN MAN

In 1824, Beckwourth joined a fur-trapping expedition in the Rocky Mountains as a horse-handler. Fur-trappers, also known as mountain men, roamed the American wilderness, hunting, laying traps and trading the pelts (skin and fur) of beavers, otters, bears and other animals. In doing this, they explored new lands and marked out routes that would be passable to wagons (trailblazing). Mountain men led hard, dangerous lives and Beckwourth encountered freezing temperatures, snowstorms and near-starvation.

> 'Being possessed with a strong desire to see the celebrated Rocky Mountains, and the great Western wilderness so much talked about, I engaged in General Ashley's Rocky Mountain Fur Company.'
>
> James Beckwourth

AMERICAN WILDERNESS

During his time in the mountains, Beckwourth maintained good relations with Native American tribes, and learnt their methods of hunting and trapping. In around 1828, he went to live with the Crow people where he traded furs with white traders. While he was there, he learned the Crow language, customs and ways of living, and married at least two Crow women.

The young Crow warrior, Pine Leaf, was thought to be one of Beckwourth's many Crow wives.

PATH TO GOLD

He returned to white settlements in 1833 and in 1850, discovered a safe route - a former Native American path - through the Sierra Nevada mountains, now called the Beckwourth Pass. When gold was found in California, he improved the trail so that settlers and gold seekers could pass along the route in wagons bound for California.

He also became a scout and guide for the US Army, including during a campaign against Native Americans. He may have witnessed the Sand Creek massacre in 1864, which saw Cheyenne men, women and children killed. He later returned to visit the Crow, where, in 1867, he died of natural causes while on a hunting trip. Some rumoured that he had been poisoned by the Crow tribe as they felt they could no longer trust him, although there is no evidence of this.

ADVENTURE STORIES

In the 1850s, Beckwourth told his life story to a journalist, resulting in a book that was published in 1856. He was known as something of a storyteller to fellow mountain men, and some accused him of exaggeration, dismissing his memoir as 'little more than campfire stories'. Instead, people favoured similarly fanciful tales of white heroes from the American West and Beckwourth's exploits were largely forgotten.

Beckwourth may not have led a blameless life, but he was one of the most influential mountain men of the American West, a former enslaved person who defied all constraints put on him and rose to make his mark on history.

The Railway Age

The advent of the railways in the 1830s revolutionised travel across the planet. Steam locomotives could transport people and goods at unprecedented speeds, taking their passengers to new and exciting places. Connecting towns, cities and ports, trains led to new settlements and stations, and whisked holidaymakers to seaside resorts and grand new hotels.

Railway termini are our gates to the glorious and the unknown. Through them we pass out into adventure and sunshine.

British author E. M. Forster

STEAM POWER

The opening of England's Liverpool and Manchester Railway in 1830 signalled the beginning of the railway age. It was the first to carry passengers and goods using only steam-powered (and not horse-drawn) locomotives. In its first year, it carried half a million passengers, along with cotton, from the port of Liverpool to the cotton mills of Manchester, stimulating the UK's Industrial Revolution.

The economic benefits of the railway soon led to its expansion across Europe, North America, Russia and the rest of the world. By the 1850s, train tracks connected nearly every country in Europe, and in the USA, some 34,000 km (21,100 miles) of railway were constructed between 1850 and 1860.

TAKE A BREAK

The cheap cost of train tickets meant people could head off on day trips and holidays, and seaside resorts and tourist attractions grew to welcome the crowds. In the 1840s, the English businessman Thomas Cook arranged cheap railway trips around England and Scotland, and eventually took groups across to Belgium, France and Germany. In June 1863, he organised the first excursion from England to Switzerland via Paris – the Alps mountains captivated travellers in the 19th century and the train cut the journey time from two weeks to two days. This was the beginning of the modern tourist industry.

RAILWAY ADVENTURES

Elsewhere, the French Riviera on the Mediterranean saw an explosion of visitors from all over Europe, with 100,000 people travelling by train to the holiday resort of Nice in 1865. In 1883, a long-distance luxury passenger train, the Orient Express, began whisking travellers from Paris to Istanbul. The 1890s saw the building of the Trans-Siberian Railway in Russia – the world's longest single railway line. Spanning an incredible 9,289 km (5,772 miles) all the way from Moscow in the west to Vladivostok in the east, it allowed people and goods to travel across the vast expanse of the world's biggest country.

PACIFIC RAILROAD

In North America, the completion of the transcontinental railroad connected the east and west coasts, resulting in a journey that took days rather than months. Goods could be easily transported across the continent, settlers could travel to the West and new towns sprung up all along the railroad. However, Native Americans were often forced to move away from areas around the railroad and tens of millions of buffalo were slaughtered. This further destroyed the Native American way of life, and many ended up living on reservations.

Before the completion of the transcontinental railroad, travelling across North America involved a dangerous six-month trek over rivers, deserts and mountains.

Isabella Bird

(1831–1904)

Isabella Bird was a free-spirited explorer who longed to escape to far-flung places, away from the stuffy confines of British society. Venturing to countries all over the world, she climbed volcanoes, trekked through jungles, rode yaks and elephants, and fell in love with a one-eyed outlaw living in America's Rocky Mountains. She was also a gifted writer and successful author of travel books.

EARLY ADVENTURES

As a child, Isabella suffered from constant ill-health. Following surgery on her spine, doctors recommended a 'change of air' so, aged 23, she set sail on her first overseas adventure and spent seven months exploring Canada and the USA. Her health improved and Isabella relished the freedom of travel, away from the strait-laced society of Britain, where women had little independence outside the family home.

TROPICAL ISLAND

In 1872, at the age of 40, Isabella set off again for the USA, stopping first in Australia, New Zealand and then Hawaii, where she stayed for eight months. She adored the tropical climate, where you could pick pineapples off trees. She also climbed two of the biggest volcanoes in the world and learnt to ride horses astride like a man and not side-saddle as many women were required to do in refined company in Europe and the USA.

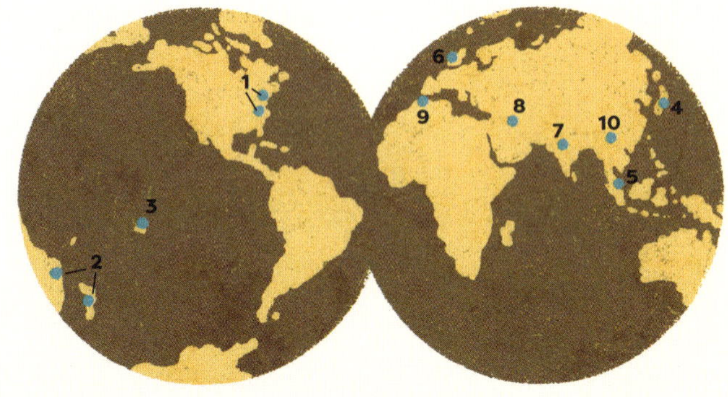

Places Isabella Bird visited included: 1. USA and Canada; 2. Australia and New Zealand; 3. Hawaii; 4. Japan; 5. Malaysia; 6. Ireland; 7. India; 8. Iran; 9. Morocco; 10. China

WILD WEST

Isabella then sailed to the USA and headed to the Rocky Mountains, where she positively revelled in the barren and hard conditions. There she met the mountain man Jim Denver, a rugged fur trapper and outlaw, who at some point in life had been mauled by a grizzly bear. He had lost one eye, and one side of his face was, as Isabella put it, 'repulsive' whereas the other side still bore his once-handsome features. She nonetheless turned down his offer of marriage and left the Rockies. A year later, Jim Denver was shot dead in a brawl.

FAR EAST

In 1878, Isabella went travelling again, this time to Asia, where she explored regions that no European had ever seen before. In Malay (now Malaysia), she rode elephants, one drenching Isabella as it showered itself. In China, she trekked up snow-capped mountains, slept in flea-infested barns and was very nearly killed by a mob who attacked her with stones, forcing her to hide in a hut. While travelling, Isabella wore Chinese-style robes, because she found it more comfortable with 'no corsets or waistbands, or constraints of any kind', and she could hide a revolver within their layers.

MEDICAL MISSIONS

After a brief marriage to an Edinburgh doctor who died in 1885, Isabella trained as a nurse and travelled to Ireland and India, where she founded two hospitals, including the Henrietta Bird Hospital for Women in Punjab. Now approaching 60, she rode horses and yaks through northern India and crossed rampaging rivers along the Tibetan border. In later years, she travelled to the Middle East and Morocco.

TRAILBLAZER

During her remarkable journeys, Isabella travelled thousands of kilometres, visiting every inhabited continent except South America. She explored many dangerous regions – sometimes setting out on her travels alone – and had countless adventures. Back in the UK, Isabella gave lectures and her books turned her into a household name. In 1892, she became the first woman to join the Royal Geographical Society – women were normally barred as it was thought that they were not well suited to being explorers! She died in 1904, although it's said her bags were packed for another trip to China.

> 'Truly a good horse, good ground to gallop on, and sunshine, make up the sum of enjoyable travelling.'
>
> Isabella Bird

Robert O'Hara Burke and William John Wills

(1821–1861 and 1834–1861)

In 1860, Robert O'Hara Burke and William John Wills famously led the first European expedition across the vast continent of Australia from the south to north. The return journey, however, met with disaster and neither explorer made it back alive. Only one member of their group, John King, survived, thanks to the help of the Aboriginal Yandruwandha people, who understood how to endure the extreme conditions of the Australian desert.

CROSSING AUSTRALIA

Most of inland Australia was unknown to European settlers in 1860. Consisting largely of desert and low mountain ranges, the inhabitants of these remote regions were Australian Aboriginal peoples who had lived there for thousands of years.

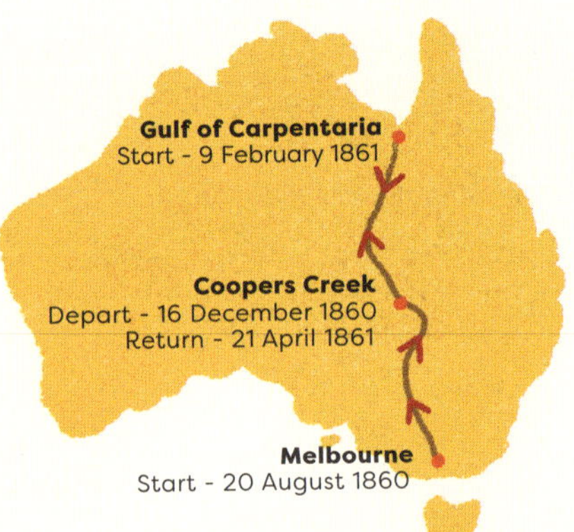

New settlers who had grown rich from mining gold were keen to fund the first European expedition to cross Australia from the south coast to north coast. They planned to map Australia's landscape and hoped the explorers would make scientific discoveries. They also wanted to find a route for an overland telegraph line and find new grazing land for European farmers.

THE GRAND EXPEDITION

Irish soldier and police officer Robert O'Hara Burke led the expedition. Second in command was Englishman William John Wills, a surveyor and astronomer. It was the most expensive expedition ever to be mounted in Australia. It involved 18 men, 25 camels, 23 horses, two years of rations, 50 gallons of rum and even an oak table. As the explorers set off from Melbourne on 20 August 1860, some 15,000 people came to watch the strange spectacle before them.

When they reached the middle of the country, Burke left the bulk of the party and supplies there and headed to a river called Cooper's Creek with a smaller group of explorers so they could move more quickly. From Cooper's Creek, Burke decided to make the final dash north with only three companions – William Wills, Charles Gray and John King – and asked the remaining party to wait for their return.

More than two months later, in February 1861, the four men finally reached the Gulf of Carpentaria in northern Australia. They were unable to make their way through thick swamps to the ocean, but they had reached their goal! On the journey home, however, tragedy struck.

THE FATAL JOURNEY HOME

Heading back south, Charles Gray died of starvation, and when the remaining men reached Cooper's Creek, they found it deserted. Unbelievably, the rest of the party had waited four months for them but had left just nine hours earlier.

By now, the three men were extremely low on supplies and were too weak to continue on to Melbourne. They decided to head to a cattle station, Mount Hopeless, but along the way, Burke and Wills – suffering from malnutrition, exhaustion and hypothermia – died.

THE SOLE SURVIVOR

John King, who nearly died himself, was saved by the Yandruwandha who accepted him into their community and looked after him until European rescue groups arrived months later.

> 'They [the native Yandruwandha people] appeared to feel great compassion for me when they understood that I was alone on the creek and gave me plenty to eat.'
>
> John King

IMPACT OF EUROPEAN EXPLORATION

The Burke and Wills expedition is one of the most famous in Australian history. However, it was to have a devastating impact on the Indigenous population. The maps and information from the expedition revealed huge areas of grazing land. This in turn led to more Aboriginal peoples being forcibly removed from their lands and paved the way for the eventual decimation of their way of life.

For thousands of years, Aboriginal peoples of Australia have navigated their way across the lands and seas using 'songlines'. Songlines describe the journey of Aboriginal ancestral spirits as they created the world (a period known as Dreamtime) and contain vital information about the landscape. Passed down the generations from elders to the young, and retold in song, stories, dance or art, they enabled Aboriginal peoples to travel vast distances and survive in the inhospitable deserts of Australia.

Songlines

SINGING THE LANDSCAPE

The Aboriginal peoples of Australia traditionally moved from place to place with the seasons or to find new hunting grounds. They also travelled to visit sacred sites, to trade or to meet other groups for ceremonies. As they had no written language, they didn't write directions down, nor did they use navigational tools such as compasses – they found their way by using songlines.

When crossing the land, travellers could repeat the words of a songline or re-enact the stories through dance. This helped them to memorise the landscape around them, and by repeating them in the correct order, they could travel great distances. The songlines would be shared by different Aboriginal clans, with each coming up with their own verse, and they criss-crossed the entire continent of Australia. Some were thousands of kilometres long!

THE SING TREE

To help point travellers in the right direction, songlines described various sacred sites or 'markers' on the landscape, such as a distinctive tree with twisted branches or a bend in the river. These acted as signposts – much like road signs that you find on the side of the road today. A well-known 'marker' is a scarred tree – a tree that has had some of its bark removed. One of the most important is a tree named Yingbeal (meaning 'sing tree') in Melbourne, which marks the place where five different songlines meet.

SONGS FOR SURVIVAL

Songlines also passed on knowledge about animals and plants and other important information needed to survive in the tough environment of the outback. They might indicate where travellers could find water, what to eat or how to get around certain features, such as forests or swamps. In a dance, kangaroo hunters might demonstrate how a kangaroo's ears move when they sense movement – information that is really useful if you want to get up close enough to hunt a kangaroo!

BROKEN LINES

When Europeans arrived in 1788 and began to drive the Aboriginal peoples out of their ancestral homes, many of the songlines were broken up. At the same time, European settlers relied on Aboriginal guides to show them the best routes – many of which were based on songlines. These then became tracks used by settlers and some later became the major highways used in Australia today. One of the longest is Stuart Highway, which runs 2,834 km (1,761 miles) from Port Augusta in the south ('where the seagulls lived') to Darwin in the north ('the crocodile's home').

'Our history was not written by people with pen and paper . . . We learned from our grandmothers and grandfathers as they showed us [our] sacred sites, told us the stories, sang and danced with us the Tjukurpa (the Dreaming Law). We remembered it all in our minds, our bodies and feet as we danced the stories.'

Nganyinytja, a Pitjantjatjara elder

African Adventures

While Europeans had explored the coast of Africa, the interior's impenetrable rivers, rainforests, swamps, not to mention the vast Sahara Desert had made further exploration difficult. All this was to change in the 18th and 19th centuries when Europeans, out of scientific curiosity, missionary zeal or commercial desire, began to venture further into Africa. Exploring the continent's mysterious interior was extremely dangerous and many succumbed to tropical diseases, such as malaria, or were killed in conflicts with Indigenous peoples.

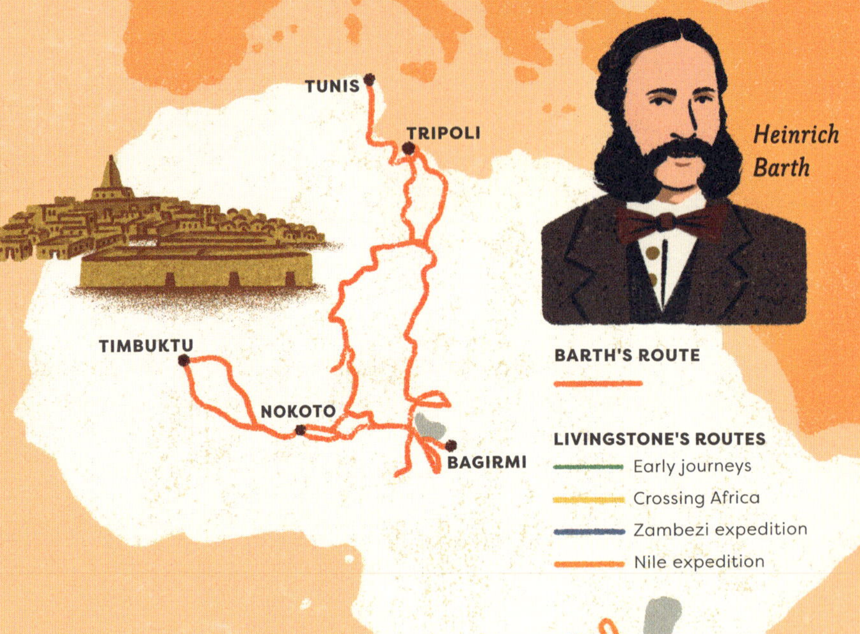

GREAT LAKES AND RIVERS

To explore the vast interior of Africa, expeditions often focused on the continent's great rivers and lakes as it was easier to travel by water than by land. Finding where the world's longest river, the Nile, began obsessed many explorers in the 1900s, whilst others explored the Zambezi and Congo Rivers in central Africa and Niger River in the west. Europeans were usually helped by local people who served as interpreters, guides, labourers and servants.

HEINRICH BARTH (1821–1865)

Unlike other explorers who wanted to map, 'discover' or make money from their African adventures, the German academic Heinrich Barth sought to better understand the richness of African culture. A self-taught Arab speaker, from 1850, he spent more than five years travelling some 19,000 km (12,000 miles) through Islamic Africa, documenting everything he could about the history, customs and languages of the people he met. On his return home, he wrote many books about his discoveries, helping to teach the West about some of Africa's amazing cultures.

DAVID LIVINGSTONE (1813–1873)

David Livingstone was one of the most famous adventurers of the 19th century. Born in Scotland in 1813, he was driven to explore Africa so he could introduce people to Christianity and free them from slavery. In the process, he travelled to many places that had never been seen by Europeans before.

Between 1849 and 1855, he walked from one side of the continent to the other, trekked across the Kalahari Desert to the great Lake Ngami, traced the path of the Zambezi River and came across one of the biggest waterfalls in the world, which he named Victoria Falls (after Britain's Queen Victoria). He faced many dangers, nearly died of thirst while crossing the Kalahari and even survived a lion attack. When he returned home, he became a national hero. In 1865, he set out on what was to be his final great expedition: the search for the source of the River Nile.

HENRY STANLEY (1841–1904)

After nothing was heard from Livingstone for many months, the explorer and journalist Henry Morton Stanley set out in 1869 to find him, promising his *New York Times* editor, 'Wherever [Livingstone] is, be sure I shall not give up the chase.' Two years later – after wading neck-deep through crocodile-infested swamps, and battling malaria, starvation and dysentery – he discovered Livingstone in poor health in the village of Ujiji, near Lake Tanganyika. There, he famously greeted him: 'Dr Livingstone, I presume.'

Livingstone died shortly afterwards. Stanley, however, went on to chart much of central Africa and played a major role in the annexation (conquest) of the Congo region by Belgium. He also helped to oversee the construction of roads in the Congo, often by brutal methods of forced labour.

'If you have men who only come if they know there's a good road, I don't want them. I want men who will come if there is no road at all.'

David Livingstone

Mary Kingsley

(1862–1900)

Mary Henrietta Kingsley once lived the typical life of a Victorian woman, keeping house and looking after her sick parents. She then made the extraordinary decision to travel on her own to West Africa, where she fought her way through jungles, waded through swamps and encountered crocodiles, leopards and gorillas. Her aim was to collect rare zoological specimens but also to better understand African culture.

DREAMS OF ADVENTURE

Like many girls in 19th-century England, Mary had little formal schooling, but she enjoyed reading her father's extensive collection of travel and science books. When her parents died, she was able to escape home life, seek adventure and travel to a part of the world that had always fascinated her – Africa.

DANGEROUS ENCOUNTERS

In 1893 and 1894, Mary travelled through many West African countries, including Sierra Leone, Angola, Nigeria and a dangerous part of Gabon that no European had ever seen. Dressed as a prim Victorian spinster, head-to-toe in black, she ventured through dangerous jungles, climbed West Africa's highest peak, Mount Cameroon, and once came face to face with a leopard as she clambered out of a forest stream. On one occasion, a huge crocodile tried to haul itself into her canoe, but she managed to escape after giving it a 'clip on the snout'!

Places visited by Mary Kingsley

GUINEA
SIERRA LEONE
NIGERIA
CAMEROON
GABON
CONGO
ANGOLA
SOUTH AFRICA

As she travelled, Mary collected rare insect, fish and plant specimens, and maintained a deep respect for African customs and society. After returning to England in 1895, she published entertaining accounts of her adventures and wrote about her opposition to many of the common European practices in Africa.

> 'I have never seen anything to equal gorillas going through the bush. It's a graceful, powerful, superbly perfect hand-trapeze performance.'
>
> Mary Henrietta Kingsley

Florence Baker

(1841–1916)

Florence Baker's story is one of extraordinary resilience. Born in Romania, she was orphaned as a child, sold into the Ottoman slave trade and as a teenager was bought by English traveller Samuel Baker. Together, they travelled to Africa, took part in the search for the source of the Nile and fought to abolish slavery.

SOLD INTO SLAVERY

At the age of 14, Florence Baker found herself on an auction block in Bulgaria, where she caught the eye of the middle-aged explorer Samuel Baker. He bought her – perhaps rescuing her from a life of slavery – and, in 1861, took her with him to Africa.

INTO AFRICA'S INTERIOR

Travelling through Sudan, they sailed up the Nile River as far as Gondokoro (in modern-day South Sudan), where they continued the journey by foot. The trip was fraught with danger as they ventured through Africa's uncharted, insect-infected wilderness. Florence proved herself an invaluable member of the party as she could speak Arabic and was able to communicate with their team of local servants when a dispute broke out. They didn't find the source of the Nile. However, they were the first Europeans to see Murchison Falls and Lake Albert in Uganda.

THE FINAL ADVENTURE

In 1865, they returned to England. Samuel was knighted by Queen Victoria, but the monarch refused to accept Florence at court. Between 1870 and 1873, the Bakers made another expedition to East Africa, where they hoped to drive out slave traders. In western Uganda, the Bakers were defeated in a battle where Florence is said to have carried in her bag rifles, brandy, two umbrellas and a pistol.

> 'I really hate the sight of them [slave traders] . . . they remind me of olden times.'
>
> Florence Baker

Gertrude Bell

(1868–1926)

Gertrude Bell turned her back on the restrictions placed on women in Victorian England to embark on an astonishingly adventurous life of exploration. She climbed mountains in the Alps and travelled across the Middle East as an archaeologist, diplomat and author. Speaking Arabic fluently, she immersed herself in the culture of the Middle East and played a key role in establishing the modern country of Iraq.

The region where Gertrude Bell worked and travelled

DEFYING CONVENTION

Gertrude was one of very few women to attend Oxford University in the 19th century and the first to earn a first-class degree in history. In 1892, she travelled to Iran to visit her uncle, sparking a life-long interest in the Middle East. Over the following years, she learnt several languages and wrote books about her travels, including one about her intrepid journey through the Syrian Desert.

> 'To wake in that desert dawn was like waking in the heart of an opal . . . See the desert on a fine morning and die – if you can!'
>
> Gertrude Bell

UNEARTHING THE PAST

Many of her travels were driven by a passion for archaeology, and her expeditions included a 1909 trek along the Euphrates River in northern Syria. In 1913, having purchased 17 camels, she undertook a dangerous 2,900-km (1,800-mile) round trip from Damascus to Ha'il and on to Baghdad. Along the way, she took photographs, made maps and examined ruins, and was held captive in Damascus by the Ottoman authorities.

MOUNTAIN PEAKS

Alongside her travels in the Middle East, Gertrude climbed mountains in the Alps, including its highest peak, Mont Blanc. One of the Alpine peaks, Gertrudspitze, was named in her honour after she became the first person to climb it in 1901. The following year, she almost lost her life climbing Finsteraarhorn, when a blizzard descended, forcing her to spend 50 hours on a rope, desperately clinging to the rockface. On another occasion, she climbed a mountain in the Alps in her underwear, as at that time, there was no suitable clothing for women mountaineers!

GERTRUDE OF ARABIA

When the First World War broke out in 1914, the British were keen to drive out the Ottoman (Turkish) Empire, which ruled much of the Middle East. The British military formed an intelligence, or spy, agency made up of people who had expert knowledge about the region, which included Gertrude Bell and T. E. Lawrence (famously known as Lawrence of Arabia). Lawrence wanted to gather support from Arab people, and Bell helped to get information about Bedouin tribes in the Middle East. She was the only female political officer in the British Army, and also drew maps to help the army cross the region safely.

Gertrude outside her tent in Babylon

CREATION OF IRAQ

The end of the war saw the collapse of the Ottoman Empire and Britain created new countries out of the land it had once controlled. Bell was asked to produce an analysis of Mesopotamia, as the region of Iraq was then known. She helped to work out the political structure of the newly formed country of Iraq, helped install its first king and worked long hours with tribal leaders to figure out the most sensible borders.

Bell then threw herself into archaeology, establishing an archaeological museum in Baghdad, with a collection that was once considered among the most important in the world.

Rediscovering Old Civilisations

Explorers have often ventured to remote regions in a quest to find traces of ancient civilisations or legendary cities 'lost' in the mists of time. Over the last two centuries, expeditions have unearthed ruins and treasures that stretch back thousands of years, from the tombs of ancient Egyptian pharaohs to an Inca city perched high in the Andes mountains.

JOHANN BURCKHARDT

In 1812, Swiss explorer Johann Burckhardt (1784–1817) was travelling through Jordan, when he heard local people talk about nearby ancient ruins. Having convinced his guide to take him there, he became the first European to see the incredible fourth-century city of Petra, its elegant buildings carved out of red sandstone cliffs. Burkhardt later travelled to Cairo and then up the Nile River, where he discovered at Abu Simbel the temple of the Egyptian pharaoh Ramesses II and its colossal statues.

Henry Mouhot at Angkor Wat

HENRI MOUHOT

In April 1858, French naturalist Henri Mouhot (1826–1861) sailed to Thailand to explore Southeast Asia. In 1860, while travelling through Cambodia, he came across the ruins of Angkor – the sprawling capital of the Khmer Empire, including its principal temple and largest religious structure in the world, Angkor Wat. While local people knew about the ruins, Mouhot made detailed reports of what he saw and alerted the West to its importance as an archaeological site.

Johann Burckhardt outside the city of Petra

HIRAM BINGHAM

Adventurers had long roamed the Peruvian countryside in search of fabled cities rumoured to be filled with gold. In July 1911, the American academic Hiram Bingham (1875–1956) was travelling through the Urubamba Valley in Peru, when a local farmer told him of some ruins at the top of a nearby mountain. He was eventually led by an 11-year-old boy to stone terraces marking the entrance to Machu Picchu. He had found no gold, but the remains of a remarkable 15th-century Inca citadel.

Hiram Bingham at Machu Picchu

Freya Stark in front of Alamut Castle

HOWARD CARTER

In November 1922, English archaeologist Howard Carter (1874–1939) was overseeing excavations of tombs in Egypt's Valley of the Kings when a boy who worked as a water fetcher came across a stone step. They discovered it led down to a secrete underground chamber. There, Carter and his patron Lord Carnarvon found an immense collection of gold and treasures and, in an innermost chamber, the sarcophagus and mummy of the pharaoh Tutankhamun. It was a stunning discovery and sparked global interest in ancient Egypt.

Howard Carter with Tutankhamun's sarcophagus

FREYA STARK

When she was a young girl, Freya Stark (1893-1983) was given a copy of *One Thousand and One Nights*, a collection of Arabic folk tales, inspiring a lifelong fascination with the Middle East. She went on to study languages at university, and later taught herself Arabic before setting off to travel around the Middle East to places where no Westerners had ever visited. She found the legendary valley of Alamut and the Valleys of the Assassins and spent two months in the remote deserts of Yemen in search of the lost ancient city of Shabwa. Often travelling alone, Stark spoke with the local people in their own language, giving her great insights into their lives. A talented writer, she wrote many popular books about her travels and continued her adventures well into her seventies.

'To awaken quite alone in a strange town is one of the pleasantest sensations in the world. You are surrounded by adventure.'

Freya Stark

The Villas-Bôas Brothers

ORLANDO (1914–2002) • CLÁUDIO (1916–1998) • LEONARDO (1918–1961)

In the early 1940s, three young brothers, Orlando, Cláudio and Leonardo Villas-Bôas, joined and went on to lead a 20-year expedition to the dense jungle of central Brazil. They spent the rest of their lives fighting to protect the Indigenous tribes of the Amazon, and their work led to the creation of the Xingu National Park, the first protected area of land in South America.

UNEXPLORED BRAZIL

The Villas-Bôas brothers were working in office jobs in São Paulo, Brazil, when they heard about a government expedition to the uncharted mountains and tropical rainforests of central Brazil. At the time, most of Brazil's population lived along or near its 4,000-km (2,500-mile) coastline and much of inland Brazil remained entirely unexplored.

DEEP IN THE RAINFOREST

The Brazilian government wanted to survey the upper reaches of the Xingu River that crossed the Mato Grosso region of central Brazil and clear it for settlement. The plan was to open up trails, explore rivers and carve out airstrips where planes could land. They knew that native populations lived deep in the rainforest, where their isolation had protected them from the deadly epidemics introduced by European settlers in the 1500s.

THIRST FOR ADVENTURE

Hungry for adventure and bored with city life, the brothers bluffed their way into the 50-man expedition in January 1944, when it reached the upper Araguaia River, despite not having the rugged frontier-type background originally required. Using machetes, they hacked their way through the dense forest, which was teeming with blood-sucking leeches, mosquitoes and venomous snakes. Progress was painfully slow as they battled intense humidity, dehydration and hunger. It took a gruelling 11 months for them to cross 320 km (200 miles), finally reaching the Xingu River in November 1945.

Women from an Indigenous tribe carrying baskets of manioc to their village

Indigenous people playing the uruá flute

Indigenous people canoe fishing

FIRST ENCOUNTERS

Over the next 20 years, the expedition opened up 1,500 km (930 miles) of trails, explored 1,000 km (620 miles) of river, established air strips and founded more than 30 towns. Along the way, they made contact with around 14 Indigenous tribes living along the Xingu River that had had no previous contact with Westerners. It fell to the Villas Bôas brothers to negotiate with the tribes and to make peace with them. Unlike many Brazilians who feared Indigenous peoples and viewed them as savages, the brothers were fascinated by the beauty and cultural richness of the people they came across and realised they had no protection from modern society, which could destroy their way of life.

PROTECTED LAND

In the 1950s, the brothers began advocating that the land in the Xingu region be made safe for the Indigenous tribes who lived there. In 1961, this became a reality when the Xingu National Park was established, protecting some 26,000 square kilometres (16,000 square miles) of land, the largest Indigenous protected area in South America and the first of its kind in the world. The Xingu National Park still exists today but is constantly under threat from deforestation and the pollution of its rivers.

> 'The true defence of the Indian is to respect him and to guarantee his existence according to his own values. Until we, the "civilized", create the proper conditions among ourselves for the future integration of the Indians, any attempt to integrate them is the same as introducing a plan for their destruction.'
>
> Orlando and Cláudio Villas-Bôas

Nellie Bly

(1864–1922)

On 14 November 1889, the American journalist Nellie Bly set sail from New Jersey, USA. Her quest was to beat the record of Phileas Fogg, the fictional hero of the famous 1873 book *Around the World in Eighty Days.* People thought a woman could never make the 42,000-km (25,000-mile) journey, but Nellie did just that, also beating another female reporter racing against her in the opposite direction.

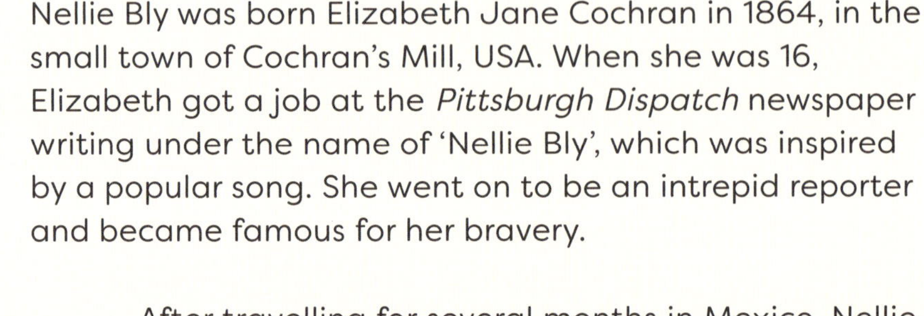

Nellie Bly was born Elizabeth Jane Cochran in 1864, in the small town of Cochran's Mill, USA. When she was 16, Elizabeth got a job at the *Pittsburgh Dispatch* newspaper writing under the name of 'Nellie Bly', which was inspired by a popular song. She went on to be an intrepid reporter and became famous for her bravery.

After travelling for several months in Mexico, Nellie worked undercover at a women's asylum so she could write about the cruel treatment of the mentally ill. In 1889, Nellie pitched the idea of travelling around the world to her editor, who responded that a woman would need to carry too much luggage and could never make the journey on her own. 'No one but a man can do this,' he said.

TRAVELLING LIGHT

Nellie went ahead anyway and bought herself a sturdy gown and a travel bag measuring just 41 by 18 cm (16 by 7 inches) into which she packed essentials for the trip. According to her account, she took with her: '. . . two travelling caps, three veils, a pair of slippers, a complete outfit of toilet articles, ink-stand, pencils, and copy-paper, pins, needles and thread, a dressing gown, a tennis blazer, a small flask and a drinking cup, several complete changes of underwear, a liberal supply of handkerchiefs and fresh ruching, and most bulky and uncompromising of all, a jar of cold cream to keep my face from the varied climates I should encounter.'

THE ADVENTURE BEGINS

The first leg of Nellie's journey via steamship to England was rough and she was horribly seasick. Undeterred, she travelled nonstop for two days by road, rail and boat to France. In the middle of the night, she then jumped on a train and raced to Brindisi, Italy, where she boarded a steamship to Egypt. She continued to the Red Sea and on to Yemen, Sri Lanka and Singapore – where the lonely traveller bought herself a fez-wearing miniature monkey, which she called McGinty.

THE BIG RACE

All the while, *The World* newspaper published daily articles on Nellie's globe-trotting trip and ran a contest asking readers to guess her return time, which attracted over a million entries. At this point, Nellie was unaware that a rival publication had sent another female journalist, Elizabeth Bisland, to compete against her. She had left New Jersey on the same day as Nellie but had travelled west not east.

From Singapore, Nellie's ship sailed through a violent monsoon to Hong Kong. She then crossed the stormy seas to Japan, where she caught a ship to San Francisco. Waiting for her was a specially chartered train, the 'Miss Nellie Bly Special', which sped her across the USA to New Jersey. She arrived on 25 January at 3.51 pm – 72 days, 6 hours, 11 minutes and 14 seconds after leaving. Nellie had not only beaten Fogg's journey by over seven days but also Elizabeth Bisland, who made it back to New Jersey five days later.

WELCOME HOME

Nellie was greeted by jubilant crowds and the race turned her into a celebrity. *The World* described her as the most 'widely talked of young women on Earth today'. Her picture went on to appear on games, toys and soaps, and a racehorse, hotel and train were even named after her.

> **'[It was] one maze of happy greetings, happy wishes, congratulation telegrams, fruit, flowers, loud cheers, wild hurrahs ... A ride worthy of a queen!'**
>
> Nellie Bly describing her train journey home

Around the World by Any Means

Aloha Wanderwell

A whole host of weird and wonderful contraptions have been used by people to travel across the world. Records have been broken on the back of bicycles – even on the cumbersome penny-farthing – whilst other adventurers have kayaked, canoed or raced around the globe in cars, driving across land faster than ever before.

HIGH-WHEELING

On 22 April 1884, Thomas Stevens set off from San Francisco, USA, riding on his penny-farthing – an early bicycle noticeable for its large front wheel. He packed his handlebar bag with socks, a spare shirt, a raincoat and a revolver – and wore no helmet despite sitting precariously high.

Thomas Stevens and his penny-farthing

Over three months, he cycled 6,000 km (3,700 miles) right across the USA, from the west coast to the east coast. He then jumped aboard a steamer ship to England, cycled across Europe and on to Iran, then India and China, ending his extraordinary 22,000-km (13,500-mile) journey in Japan, on 17 December 1886, becoming the first person to tour the world while riding a bicycle.

RIDING IN HER BLOOMERS

The first woman to tour the world by bicycle was the Latvian born Annie Kopchovsky, better known as Annie Londonderry. According to some accounts, two men had bet that no woman could beat Thomas Stevens's around-the-world record. Despite never having cycled before, Annie took on the challenge, hoping to win US$10,000 (about US$300,000 today) if she succeeded.

Annie Londonderry on her bicycle

THE WORLD'S MOST WIDELY TRAVELLED GAL

Canadian Aloha Wanderwell (born Idris Galcia Hall) was only 16 when she embarked on a daring around-the-world tour. Responding to a newspaper advert for 'Lucky Young Woman . . . Wanted to join an expedition!', Aloha left France in October 1922 in a fleet of Ford Model T cars, along with her travel companions and a pet monkey named Chango. Over the next five years, she travelled more than 120,000 km (75,000 miles) through 43 countries, camped at the foot of the Great Sphinx in Egypt and nearly died of thirst in the Sudanese desert.

THE AUTOMOBILE

The invention of the car had a huge impact on exploration, as people had the freedom to travel independently. History credits Karl Benz, a German engineer, with inventing the automobile in around 1885. To draw attention to the invention, Benz's wife and business partner, Bertha, decided to drive the three-wheeler automobile 106 km (66 miles) from Mannheim to Pforzheim in Germany. It was the first time an automobile was driven over a long distance.

The whole world was out there . . . I, reaching for it. The world reaching for me.

Aloha Wanderwell

Aloha Wanderwell

Thomas Stevens

Sarah Outen

Annie Londonderry

After two short cycling lessons, Annie set off from Boston, USA, on 25 June 1894. Progress was slow, as her bicycle was heavy and her long skirt and corset cumbersome. She eventually swapped her bicycle for a lighter one and ditched her skirt in favour of bloomers. Fifteen months later, she arrived back in the USA and collected her prize money.

Sarah Outen paddling in her kayak

PEDAL, PADDLE AND ROW

On 3 November 2015, the British adventurer Sarah Outen completed a solo round-the-world journey by rowing boat, bicycle and kayak. The mammoth 30,000-km (20,000-mile) expedition saw her cycling across Europe and Asia, rowing across the North Pacific (the first woman to do so), cycling across the USA and Canada, before rowing across the North Atlantic to the UK. During the four-and-a-half-year journey, Sarah had to navigate unchartered waters, and survive storms, hurricanes, dangerous currents and huge waves that at one point tore off the rudder of her rowing boat!

EXPLORERS
Over Ice and Snow

A frozen wilderness, whether it's shifting pack ice in the Arctic or the summit of Mount Everest, represents one of the toughest challenges for explorers. Trekking in sub-zero temperatures means battling against blizzards, frostbite or, if at the top of a mountain, dangerously low levels of oxygen. Adventurers, however, like a challenge and have endeavoured to cross frozen seas and scale the world's highest mountains, embarking upon some of the most epic journeys of exploration.

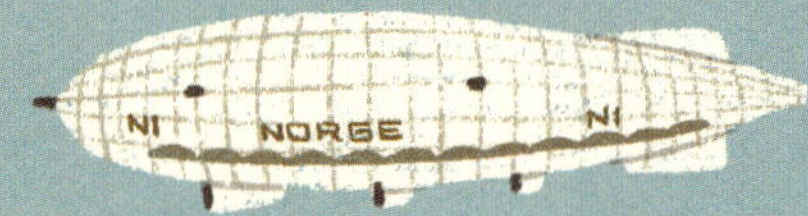

For more than 400 years, explorers risked their lives searching for the Northwest Passage, a sea route north of Canada, connecting the Atlantic and Pacific Oceans. Countless men died trying to cross its treacherous waters, navigating around huge icebergs, only to be trapped in its frozen wastes for months on end.

The Search for the Northwest Passage

THE SEARCH IS ON

European seafarers first began searching for the Northwest Passage in the 1400s, as they were keen to find a shorter, westward route to the trade markets of Asia. In the 1490s, Christopher Columbus and John Cabot sailed west looking for a sea route to Asia but instead found America. Finding the Northwest Passage would be a way around this continent.

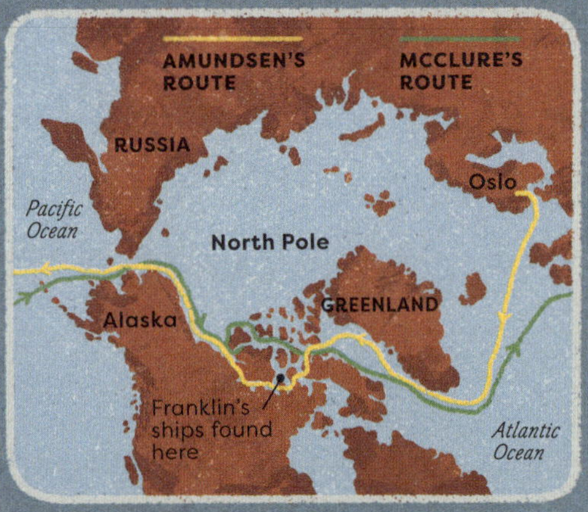

In the 1500s, Spanish and French expeditions tried and failed to find the Northwest Passage, although Jacques Cartier (p21) managed to sail as far as Quebec in Canada. English seamen also sailed in search of a route, including Martin Frobisher, who made three voyages to the Arctic in the 1570s with no success. In 1610, a Dutch expedition led by English explorer Henry Hudson ended up in what is now called Hudson Bay off northeast Canada where his ship became trapped in ice. The crew eventually mutinied and set Hudson and eight others adrift in a small boat, never to be seen again.

FATAL EXPEDITION

One of the most famous attempts to find the Northwest Passage was led by British Navy officer Sir John Franklin in 1845. He sailed with 128 men on two ships, HMS *Terror* and HMS *Erebus*, both of which vanished. A note found later revealed that the ships had been trapped in the ice for many months, supplies ran out and many of the crew starved to death. Some of the survivors set out on foot but were never heard from again.

SHIP AND SLEDGE

Over the subsequent decades, many other expeditions set out to find Franklin, including Irish Arctic explorer Robert McClure between 1850 and 1854. His ship was trapped in ice for three winters – by which time the crew were dying from starvation. They were eventually found by rescuers who took them by sledge back to their ship on the other side of the passage. In doing so, McClure and his crew became the first people to cross the Northwest Passage by ship and sledge.

NORWEGIAN SUCCESS

The first person to make it through the Northwest Passage solely by ship was the Norwegian explorer Roald Amundsen between 1903 and 1906. Setting out from Oslo in 1903, he used a much smaller vessel with a crew of just six. The expedition anchored on the east coast of King William Island for two years, where they conducted scientific observations and learned from the local Inuit people. They learned how to make igloo ice shelters to keep them warm and dry in the freezing Arctic conditions, how to pull their heavily laden sledges across the snow more efficiently and what clothes to wear. They set sail again on 13 August 1905 towards the Bering Strait. When Amundsen saw a whaling ship from San Francisco coming in the opposite direction he knew he had finally crossed the Northwest Passage.

Amundsen and his team wore clothing given to him by the local Inuits to keep them warm.

'The Northwest Passage was done. My boyhood dream – at that moment it was accomplished.'

Roald Amundsen

GLOBAL WARMING

Only a small ship could have followed Amundsen's route, as the water was as shallow as 1 m (3 ft) in places. It wasn't until summer 2007, when climate change caused the passage to be completely free of ice for the first time in recorded history, that larger ships could get through.

Matthew Henson and Robert Peary

MATTHEW HENSON

(1866–1955 *and* 1856–1920)

Reaching the North Pole, the most northern point on Earth, is an immense challenge for explorers. Unlike the South Pole, which is on land, the North Pole sits on constantly shifting ice in the middle of the Arctic Ocean. Matthew Henson, an African American explorer from Maryland, USA, may have been the first person to reach the North Pole, on an expedition led by US naval officer Robert Peary.

ADVENTURES AROUND THE WORLD

The descendant of enslaved people, orphan Matthew Henson was only around 12 years old when he joined the crew of a ship and sailed around the globe. Having discovered a love of travel and adventure, in 1887, he accompanied US explorer Robert Peary on an expedition to survey the jungles of South America. During this trip, Peary told Henson of his dream to become the first person to reach the North Pole.

ARCTIC EXPLORATION

The pair then made seven expeditions to the frozen wilderness of the Arctic. They faced many dangers, from snow blizzards and near starvation to the constant threat of frostbite. To survive the incredibly hostile environment, Peary hired local Inuit women to make their fur clothing, and Henson found out from them how to hunt, drive sledges, and build igloos out of snow, and learned the Inuit language.

FIELDS OF ICE

In 1908, Peary and Henson made their eighth and final attempt to reach the North Pole, travelling by ship to Ellesmere Island, Canada's most northerly island. The following March, they set out with 23 men, 19 sledges and 133 dogs. They crossed a huge field of ice, which could have cracked at any time, and slept in igloos. One by one, members of the party began turning back, leaving only Peary, Henson and four Inuit men, Ootah, Seegloo, Egingwah and Ooqueah.

NORTHERN POINT

Three days before they reached the North Pole, Henson fell into icy waters. Ootah swiftly pulled him out – seconds later and he would have died. Finally, on 6 April, Peary estimated they had reached the North Pole.

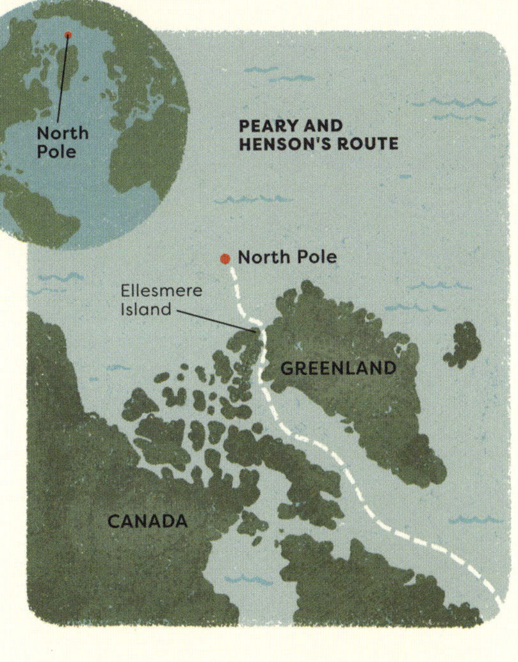

When they returned to America, Peary was given the credit for 'discovering' the North Pole. Henson, however, claimed he had reached the North Pole earlier, as part of a leading party, and his footprints were still in the snow when they returned. Despite this, Peary's account was widely accepted, although doubts have since been raised about the accuracy of his navigational records and whether he really did reach the North Pole. Regardless of whether either of them got there, both men were brave and gifted explorers and Henson's resourcefulness, sled dog-handling skills and fluency in the Inuit language played a key part in the expedition's success.

'I was in the lead that overshot the mark by a couple of miles . . . We went back then and I could see that my footprints were the first at the spot.'

Matthew Henson

OTHER MISSIONS

Later attempts were made to fly aircraft to the North Pole. In May 1926, US naval officer Richard E. Byrd and pilot Floyd Bennett declared they had made the 2,470-km (1,535-mile) journey to the North Pole from Spitsbergen in Norway, although this is now doubted. Just three days later, on 12 May, a joint Norwegian-Italian-American expedition led by the famous Arctic explorer Roald Amundsen flew over the North Pole in an airship, dropping the three countries' flags onto the icy wastes below.

The Race to the South Pole

Covered in ice up to 2 km (3 miles) deep and one of the most inhospitable places on Earth, Antarctica was the last continent to be discovered by human beings. Once seafarers established that the frozen continent existed, explorers then raced to reach parts of it no one had ever seen, including the most southern point on the Earth and the greatest geographical prize of the time, the South Pole.

ICY SHORES

People had long believed that a vast continent lay in the very south of the globe. In the 1770s, British explorer Captain James Cook concluded that a polar continent was 'probable', having seen great fields of floating ice some 120 km (75 miles) from its coast. In January 1820, Russian naval officer Fabian Gottlieb von Bellingshausen reported he had seen Antarctica's icy shores, and a British navy officer, Edward Bransfield, described seeing mountains covered with snow. American sealer Nathaniel Palmer sighted land later that year, and subsequent British, French and American expeditions established that Antarctica was a continent.

Fabian Gottlieb von Bellingshausen was the first person to see the land of Antarctica.

SCIENTIFIC EXPEDITIONS

In the late 1800s, scientific expeditions made their way to the Antarctic, including the Belgian ship *Belgica*. It was the first to winter there between 1898 and 1899, when it became trapped in ice. The crew narrowly escaped scurvy by eating fresh penguin and seal meat, a tip that would prove invaluable for crewmember Roald Amundsen, who would later reach the South Pole. In 1899, a British expedition led by Carsten Borchgrevink spent the winter on the mainland. They were the first to set up a shore base on the mainland and the first to use dogs and sledges.

DISCOVERY

The first attempt to find a route to the geographic South Pole was led by British explorer Captain Robert Falcon Scott in 1901 on the ship *Discovery*. Scott was joined by Arctic explorers Ernest Shackleton and Edward Wilson. Their principal aim was to march to the South Pole, although the expedition also aimed to map unknown coastline and collect scientific data and specimens. In fact, the three-year expedition discovered 500 new marine animals and took the first photograph of an emperor penguin chick.

SNOW BLINDNESS

On 2 November 1902, Scott, Wilson and Shackleton embarked on a harrowing overland trek towards the South Pole in temperatures as low as -45°C (-49°F). Suffering from frostbite, snow blindness and the first signs of scurvy, they were forced to turn back on 30 December. The three men made it to their ship a month later, virtually unrecognisable with long beards, swollen, peeled skin and blood-shot eyes. *Discovery*, which had been frozen solid in the ice, was freed from her icy prison with the use of explosives.

Shackleton, Scott and Wilson

Shackleton was sent home ill, but he vowed to come back, and between 1907 and 1909 led a second British expedition to Antarctica. They came within 156 km (97 miles) of the South Pole, but their food supplies ran dangerously low, and they had to turn back. The race was still on to reach the South Pole.

'I seemed to vow to myself that someday I would go to the region of ice and snow and go on till I came to one of the poles of the Earth, the end of the axis upon which this great round ball turns.'

Ernest Shackleton

Robert Falcon Scott and Roald Amundsen

(1868–1912 *and* 1872–1928)

ROBERT FALCON SCOTT

ROALD AMUNDSEN

In 1911, Britain's Robert Falcon Scott and Norway's Roald Amundsen both launched expeditions to reach the South Pole. The race would end in victory for Amundsen and tragedy for Scott and his four companions. In the brutal environment of Antarctica, even the smallest decision can be the difference between life and death.

THE RACE IS ON

Both Scott and Amundsen were determined to be the first people to reach the South Pole, although Scott also planned to conduct scientific work on his expedition. Amundsen's ship the *Fram* landed 100 km (60 miles) closer to the South Pole than the British ship *Terra Nova*, although the Norwegians didn't know for sure their route was passable.

REACHING THE POLE

Amundsen set out with four men, four sledges and 52 dogs on 18 October 1911 and arrived at the South Pole on 14 December 1911. Scott left his base camp three weeks after Amundsen, with four men, ten ponies, teams of dogs and two motorised sledges. When Scott's party reached the South Pole on 17 January 1912, they were devastated to see the Norwegian flag already there.

'. . . this is the greatest factor – the way in which the expedition is equipped – the way in which every difficulty is foreseen, and precautions taken for meeting or avoiding it.'

Roald Amundsen

DISASTER STRIKES

The British explorers then trekked back over the ice, hauling sledges and battling against -40°C (-40°F) temperatures. One by one, the men perished, first Edgar Evans, then Lawrence Oates, who with crippling frostbite and barely able to walk, bravely sacrificed himself to a snowstorm so he wouldn't hold the others up. Scott and the remaining men, Edward Wilson and Henry Bowers, were eventually trapped in their tent by a blizzard, where, weak from exhaustion, hunger and extreme cold, they too died. They were just 19 km (12 miles) from a supply depot.

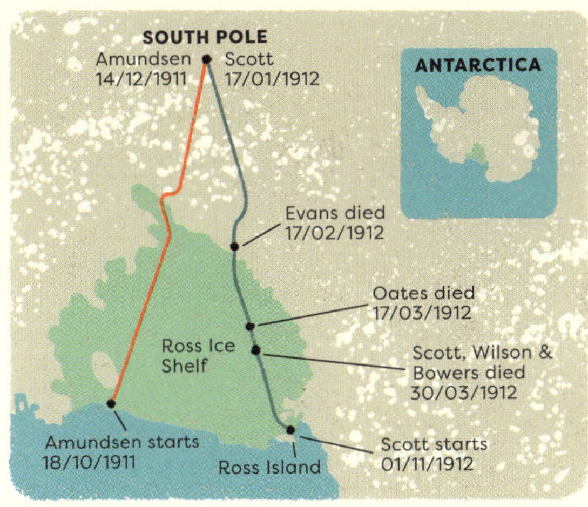

EXTREME SURVIVAL

By contrast, all of Amundsen's team made it safely back to their base camp, an achievement based largely on their experience of polar weather. Norwegians are used to cold climates and the team were experienced skiers, unlike the British team who struggled on the icy terrains. During previous expeditions to the Arctic, Amundsen had also learnt a great deal from Inuit tribes who lived there. The Norwegians wore wolf-skin suits, adapted from those worn by Inuit people, and wore their clothing loosely to reduce sweating. The Inuit had also shown Amundsen how to build igloos and design sledges.

MAN HAULING

Dogs were used throughout the Norwegian expedition and expert dog-handlers looked after them. The British planned to use motorised sledges, ponies and dogs for the first part of the journey and then continue by foot, hauling heavy sledges weighed down by rock samples. The motorised sledges broke down almost immediately, the ponies sank to their bellies in deep snow and died, and the dogs were sent back. Hauling sledges up and over glaciers was slow and exhausting work for the men.

FOOD EQUALS ENERGY

With little to eat, it is likely that Scott's team were also suffering from scurvy. Amundsen's team had eaten fresh seal and penguin meat – as the Inuit people do – as well as dog meat during the expedition. The British overcooked their seal and penguin meat, which destroyed much-needed vitamins, refused to eat their dogs (although they did eat their ponies) and ate a less nutritious biscuit as they trekked across frozen Antarctica.

> 'Had we lived, I should have had a tale to tell of the hardihood, endurance and courage of my companions which would have stirred the heart of every Englishman.'
>
> Robert Falcon Scott

Ernest Shackleton

(1874–1922)

›››››››››››››››››››››››››

Ernest Shackleton's attempt to cross the vast continent of Antarctica resulted in one of the greatest adventure stories of all time. The epic journey of survival began when Shackleton's ship was trapped and crushed by ice. Shackleton then led his crew on a seemingly impossible journey across wild seas, icy mountains and frozen wastes for more than 1,000 km (600 miles).

FROM COAST TO COAST

Having attempted to reach the South Pole on two previous expeditions, the Anglo-Irish explorer Ernest Shackleton embarked on his next Antarctic challenge: to cross the continent from coast to coast. The 3,300-km (2,000-mile) journey would be the first crossing of the uncharted wilderness from the Weddell Sea to the Ross Sea, via the South Pole.

FROZEN SOLID

Shackleton's ship *Endurance* and its 28-man crew left the island of South Georgia on 5 December 1914. Reaching the Weddell Sea, the ship slowly worked its way through floating sea ice until a dramatic drop in temperature hardened the ice around the ship, trapping it for ten long months. During this time, the crew survived by eating the rations they had brought with them and heading out on to the ice to look for penguins and seals to feed their dogs and themselves. To keep their spirits up, they played chess and football and hockey on the sea ice. The weather, however, was bitter and they endured blizzards and the long dark days of an Antarctic winter.

STRANDED ON ICE FLOES

On 21 November 1915, *Endurance* was eventually crushed by shifting ice and sank, leaving 28 men now isolated on drifting ice with limited supplies, no means of communication and no ship. They camped for five months before Shackleton ordered the crew into three small lifeboats and they set off for the nearest land. After a harrowing six days on the open ocean, battling freezing spray and seasickness, the exhausted men landed on the barren Elephant Island.

WILD SEAS AND MOUNTAINS

Nine days later, Shackleton and five others set out in a lifeboat bound for a whaling station on South Georgia, around 1,300 km (800 nautical miles) away. For 16 days, their tiny boat was thrown about in terrifying waves and icy winds. Arriving in South Georgia, Shackleton and two of the men then made a dangerous trek over uncharted mountains and glaciers, finally staggering into a whaling station on 20 May 1916.

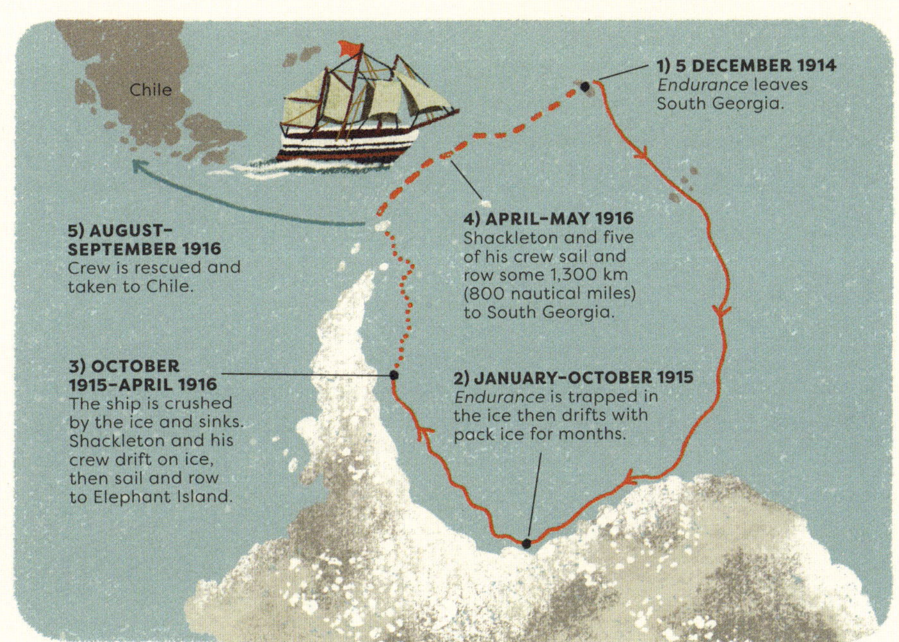

EXTRAORDINARY RESCUE

They were an alarming sight, with stringy hair and beards. Shackleton immediately sent a boat to pick up the other men on the other side of South Georgia and then set to work rescuing those stranded on Elephant Island. His first three attempts were blocked by sea ice, but they finally managed to get a small sea tug to the island on 30 August 1916. Against all the odds, every one of the *Endurance* crew made it back alive – a feat of almost superhuman perseverance and courage, not to mention heroic leadership from Ernest Shackleton.

> 'To be brave cheerily, to be patient with a glad heart, to stand the agonies of thirst with laughter and song, to walk beside death for months and never be sad – that's the spirit that makes courage worth having.'
>
> Ernest Shackleton

Ranulph Fiennes

(1944–)

From 1979–1982, the British adventurer Ranulph Fiennes led the first circumnavigation of the world via the North and South Poles. Fiennes' wife Ginny came up with the idea of the 84,000-km (52,000-mile) epic journey. Using surface transport only, they zipped across Antarctica by snowmobile, the Sahara Desert by Land Rover, and the Northwest Passage in an open motorboat.

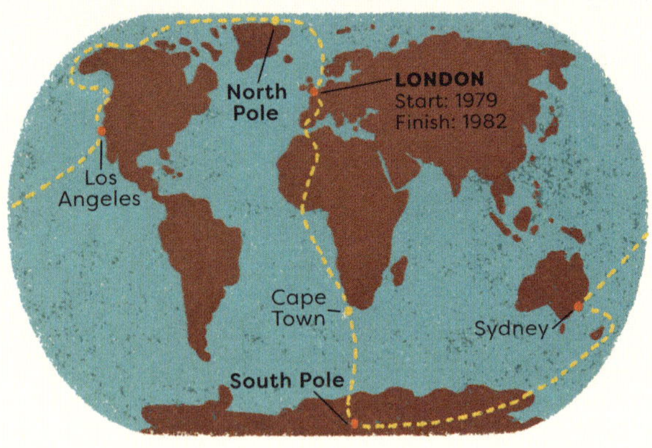

ENDS OF THE EARTH

After seven years of planning, the expedition, made up of Fiennes, Charles Burton, Oliver Shepard and a support team, left London on 2 September 1979. Heading south, they crossed Europe and the Sahara Desert in a Land Rover, then travelled by sea to Antarctica, where they wintered for eight months living in huts made of thin cardboard. On 28 October 1980, they headed to the South Pole on snowmobile, reaching it on 15 December. After playing cricket, they continued across icy crevasses and glaciers to the west coast. It took them 67 days – the fastest ever crossing of the continent.

NORTH BOUND

Over the next several months, they voyaged northwards by ship across the Pacific Ocean to Alaska, then navigated the hazardous Northwest Passage in an open motorboat, before setting out for the North Pole in mid-February 1982. After an arduous trek by snowmobile and sledge, they made it to the North Pole on 11 April. They were then forced to spend three months on a drifting ice floe, before they sailed back to Britain, having achieved their record-breaking feat.

> 'Fear must not be allowed to take hold in the first place. If you are in a canoe, never listen to the roar of the rapid ahead before you let go of the riverbank, just do it!'
>
> Ranulph Fiennes

Ann Bancroft

(1955–)

>>>>>>>>>>>>>>>>>>>>>>>>

Ann Bancroft is one of the world's greatest polar adventurers. She was the first woman to reach the North and South Poles, and she and Norwegian explorer Liv Arnesen were the first women to cross Antarctica. She is also passionate about education and inspiring women and girls to live their dreams.

DREAM CHALLENGE

Ann, who was born in rural Minnesota in the USA, has always had a passion for the outdoor world. Since the age of eight, she also longed to explore the far reaches of the world after reading about the great polar explorer Ernest Shackleton.

In 1986, Ann left her teaching job and decided to pursue her childhood dream by joining an expedition to the Arctic Pole. She travelled with seven men, and described the trek using dog sleds over shifting ice and in freezing temperatures as tough. But every day, she felt truly alive and excited, especially as they neared the North Pole.

ALL-WOMEN EXPEDITIONS

In 1992, Ann led the first team of women to ski across Greenland, and, in November of that year, led an all-woman expedition across Antarctica to become the first women's team to reach the South Pole. In 2001, Ann returned to Antarctica with Norwegian explorer Liv Arnesen, and they became the first women to complete the crossing of the continent, pulling their own sleds, sometimes with sails attached. More than three million children tracked their progress via satellite. Ann has always been passionate about children's education and established a foundation that encourages girls to follow their dreams.

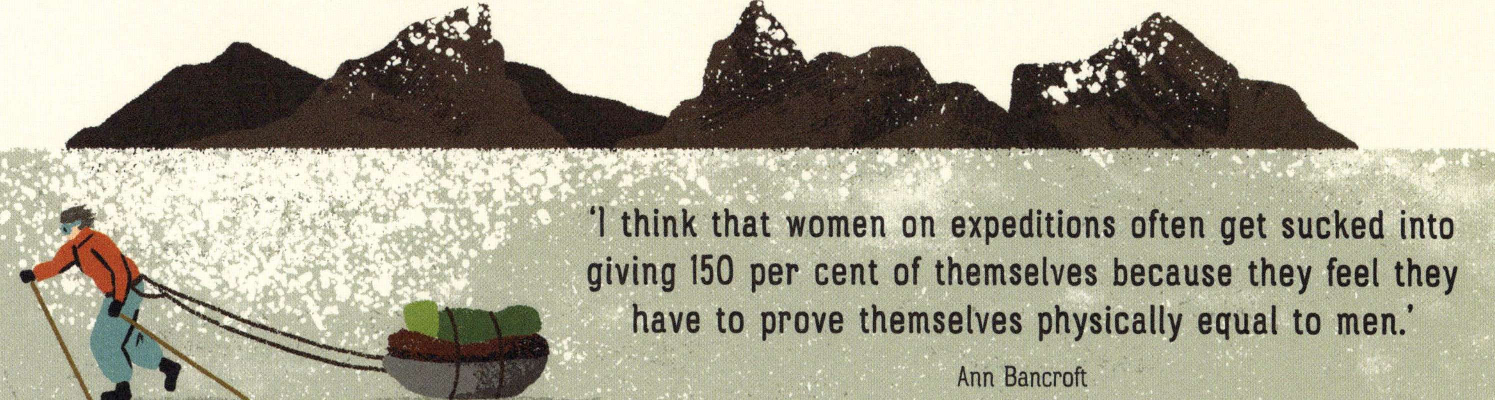

'I think that women on expeditions often get sucked into giving 150 per cent of themselves because they feel they have to prove themselves physically equal to men.'

Ann Bancroft

Mountain Climbers

Many people are driven to climb mountains and reach the highest summits around the world. Mountaineers must brave terrifying precipices, icy crevasses, crumbling rockfaces, avalanches and the dangerous effect of high altitude. Some attempt to climb the biggest and most difficult mountains across the world, pushing their bodies to the very limits of endurance.

MONT BLANC

Throughout history, people have climbed mountains, perhaps to hunt, view nearby lands, build temples or make scientific observations. From the mid-18th century, more people were lured simply by the thrill of achievement it gave them. In 1761, Swiss scientist Horace-Bénédict de Saussure was so in awe of Mont Blanc, the tallest peak in Western Europe, that he offered prize money to the first person who scaled it. Twenty-five years later, in 1786, the money was finally claimed by French doctor Michel-Gabriel Paccard and his porter, Jacques Balmat.

MOUNT KILIMANJARO

Mount Kilimanjaro, the tallest mountain in Africa, was first summitted in 1889 by a team led by German geographer Hans Meyer. Sheila MacDonald (left) from the UK was the first woman to climb Kilimanjaro, in 1927, leaving behind two male companions who had given up through exhaustion. Almost half of those who climb Kilimanjaro suffer from altitude sickness, caused by lack of oxygen, which can lead to swelling of the brain and even death.

MOTHER GODDESS

The highest mountain on Earth is Mount Everest, located in the Himalayas in a region between Nepal and Tibet. Its Tibetan name is Qomolangma, meaning 'mother goddess of the world'. The snow and ice on the mountain create deadly hazards, such as avalanches, temperatures can drop to -62°C (-80°F) and winds can blow more than 320 kph (200 mph). Low oxygen levels near the summit, meanwhile, can impair climbers' judgement, and they can experience heart attacks and strokes.

The first recorded people to climb Everest were New Zealander Edmund Hillary and his Nepalese guide Tenzing Norgay in 1953 (p102). In 1975, Japanese mountaineer Junko Tabei and her all-female team were the first women to reach the summit of Everest, at a time when women faced discrimination, and men had even refused to climb with her. Tabei was often told that Everest was 'no place for a woman' and that she should stay at home and look after her children.

Junko Tabei went on to defy expectations and complete another 44 all-female mountaineering expeditions around the world.

'I didn't intend to be the first woman on Everest. I just simply climbed a mountain.'

Junko Tabei

MOUNTAIN RECORD-BREAKERS

Seven Summits

The Seven Summits are the highest mountains found in each continent of the world. On 14 May 2018, the Australian mountaineer Steven Plain climbed all seven mountains, as well as Australia's Mount Kosciuszko, in 117 days 6 hours and 50 minutes, the fastest time ever. In 1992, Japanese mountaineer Junko Tabei, the first woman to conquer Everest, also became the first woman to climb the Seven Summits.

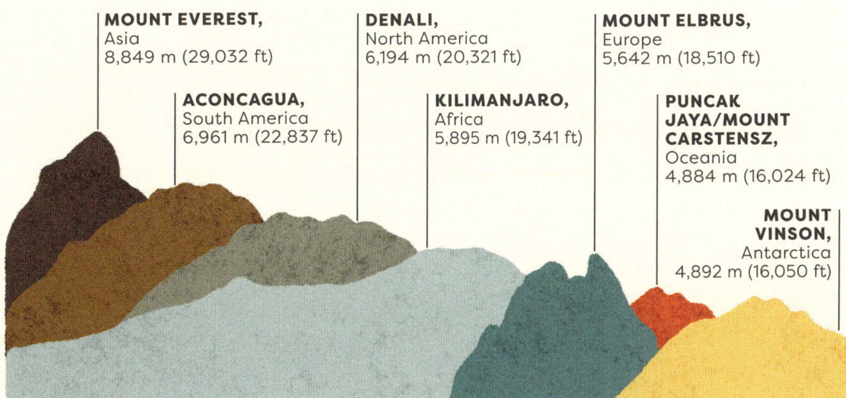

The Big 14

Another challenge for mountaineers is the 'eight-thousanders' – 14 mountains that are more than 8,000 m (26,247 ft) above sea level. All are in the Himalayan and Karakoram mountain ranges in Asia and take climbers to an extremely high altitude, nicknamed the 'death zone', because oxygen levels are dangerously low. The most treacherous is Annapurna in Nepal, which has claimed the lives of one out of every three people who has tried to reach its summit.

The first person to climb the eight-thousanders was Italian Reinhold Messner between 1970 and 1986. He scaled Mount Everest without the use of supplementary oxygen, climbed K2, the world's second-highest mountain, and claimed to have seen a yeti in the Himalayas. In 2010, Austrian mountaineer Gerlinde Kaltenbrunner became the first woman to climb all 14 mountains without oxygen.

Reinhold Messner and Gerlinde Kaltenbrunner

Edmund Hillary and Tenzing Norgay

(1919–2008 and 1914–1986)

At 11.30 am on 29 May 1953, Tenzing Norgay of Nepal and Edmund Hillary of New Zealand became the first explorers to reach the summit of Mount Everest, the highest mountain in the world. Despite being warned they would never survive the climb, they pushed on and for a glorious 15 minutes, they stood at the highest point on Earth.

UNCLIMBABLE

At 8,849 m (29,032 ft), many considered Mount Everest unclimbable. Would-be climbers have to brave freezing temperatures, deep crevasses, sheer rockfaces, not to mention the effects of extreme high altitude. Before Hillary and Norgay, many others had attempted (and failed) to reach the summit, most famously the British climbers George Leigh Mallory and Andrew Irvine in 1924. They were last seen 275 m (900 ft) short of the top before they disappeared, presumably having fallen to their deaths.

EXPERT MOUNTAINEERS

In 1952, a Swiss expedition which included Sherpa Tenzing Norgay almost reached the summit, but they had to turn back as they had run out of supplies. Many expeditions hired Sherpas, a previously nomadic people who live near Mount Everest, to help them carry their food and supplies up the mountain. The Sherpas are expert mountaineers who can adapt quickly to high altitudes. Tenzing Norgay was a highly experienced climber who had taken part in more Everest expeditions than any other person.

> 'I needed to go . . . the pull of Everest was stronger for me than any force on Earth.'
>
> Tenzing Norgay

TOWARDS THE SUMMIT

The British Everest Expedition of 1953 selected experienced climbers for its large team, including Tenzing Norgay. The expedition also enlisted mountaineer Edmund Hillary, who worked as a beekeeper when not climbing mountains. After months of planning, the expedition began to push its way up the mountain in April and May 1953, making a series of camps along the way, which are still used by climbers today.

By 26 May 1953, two men in the expedition attempted to reach the summit but, at just 100 m (330 ft) below the summit, they were forced to turn back due to bad weather and a problem with their oxygen tanks. Three days later, Norgay and Hillary prepared themselves for a push towards the summit. After a bitterly cold and sleepless night, Hillary discovered that his boots had frozen solid outside the tent and had to spend two hours defrosting them over a stove.

FINAL OBSTACLE

They finally set off at 6.30 am, and about an hour later, came across a steep rockface. Hillary wedged himself in a crack and managed to inch himself up using an ice axe, before then throwing down a rope to Norgay so he could follow. At 11.30 am, the two men reached the summit, where they shook hands before Tenzing clasped Hillary and hugged him in celebration. They spent 15 minutes taking photographs and admiring the view, ate some mint cake, and Tenzing, as a devout Buddhist, left a food offering.

'A few more whacks with my ice-axe and Tenzing and I stood on top of Everest.'

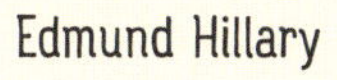

They then began their descent and returned to Kathmandu, the capital of Nepal, a few days later. News of their epic climb soon spread around the world, and they were hailed as heroes.

Air and Space EXPLORERS

When explorers took to the skies, they were able to travel further and faster than ever before. Hot-air balloons and gliders led to engine-powered aeroplanes and spacecraft that could hurtle through the Solar System towards the planet Mars and beyond. Space exploration gave us an entirely new perspective of Earth and will continue to transform our understanding of the Universe and our place in it.

Fearless Fliers

For thousands of years, humans have dreamed of taking to the skies. That desire to fly led from kite-flying in ancient China to hot-air balloons in the 1700s. Brave pioneers then experimented with gliders and steam-powered flying machines. In the late 1800s, enormous airships took flight and, in 1903, American brothers Orville and Wilbur Wright flew the world's first successful motor-operated aeroplane.

GO FLY A KITE

Since antiquity, there have been stories of people attempting to fly by strapping makeshift wings to themselves and flapping them like birds. Invariably, the attempt would end in disaster. From around 400 BCE, the Chinese built sophisticated kites, some of which may have been able to carry a person. Over the centuries, people developed ideas for contraptions that could fly, including the 15th-century genius Leonardo da Vinci, who sketched many ideas for flying machines.

Chinese kite

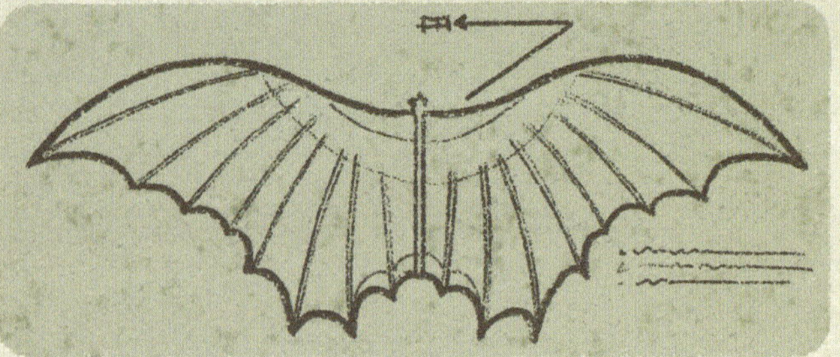

Da Vinci's sketch of wings

Otto Lilienthal's glider

HOT-AIR BALLOONS

The 1700s saw developments in flight, culminating in 1783, when French paper manufacturers Joseph and Etienne Montgolfier displayed their new invention to the public. Powered by hot air from a fire, the colourful silk balloon's first passengers were a sheep, a duck and a rooster. On 21 November 1783, the balloon then carried two men over Paris for 8 km (5 miles). Ten days later, another French scientist, Jacques Charles, travelled 40 km (25 miles) over 2 hours in a balloon filled with hydrogen.

The Montgolfier hot-air balloon

GLIDING THROUGH THE AIR

The 1800s saw English philosopher Sir George Cayley identify the main forces of flight and design many versions of gliders, which are a type of aircraft without motors. In 1891, German engineer Otto Lilienthal designed the first glider that could carry a person and fly long distances. He made more than 2,000 flights in gliders until he was killed in 1896 when a sudden strong wind caused him to lose control and crash to the ground.

WRIGHT FLYER 1

The 1800s also saw largely unsuccessful experiments with aircraft fitted with steam-powered engines, which were often too heavy or ran out of fuel quickly. From 1898, two American brothers and former bike manufacturers, Orville and Wilbur Wright, built and tested a number of kites and gliders. Later, they built an aeroplane called the *Wright Flyer 1* with a gasoline engine and two propellors.

FIRST FLIGHT

On 17 December 1903, taking off from Kitty Hawk, North Carolina, the brothers took turns flying their plane, with the first flight lasting 12 seconds, the longest 59 seconds. They had flown the world's first successful engine-powered aeroplane! This inspired others to design and develop new aircraft and modern aviation was born.

The Wright Flyer

'It was only a flight of 12 seconds, and it was an uncertain, wavy, creeping sort of flight . . . but it was a real flight at last and not a glide.'

Orville Wright

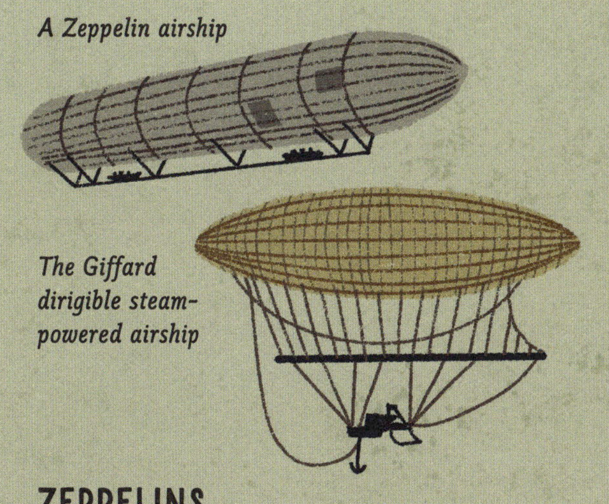
A Zeppelin airship

The Giffard dirigible steam-powered airship

ZEPPELINS

Half a century before the Wright brothers took to the skies, French engineer Henri Giffard flew a steam-powered airship more than 27 km (17 miles). Later, German inventor Ferdinand von Zeppelin developed a much larger airship, which took its first flight in July 1900. Zeppelin airships began to carry fare-paying passengers from 1910 and were used by the German military in the First World War for scouting and bombing raids.

Louis Blériot

(1872–1936)

>>>>>>>>>>>>>>>>>>>>>>>>

The early days of aviation gripped the popular imagination, as was the case on 25 July 1909 when, in a widely publicised event, two competitors attempted to fly across the English Channel for the first time. French pilot Louis Blériot ultimately won the day, earning himself worldwide fame and a £1,000 cash prize (roughly £120,000 in today's money).

Born in 1872, Louis Blériot had made a fortune in the car industry before turning to his great passion: flying. He designed and tested a series of aeroplanes until finally hitting upon a type of monoplane (a plane with one pair of wings) that would take him across the English Channel.

'I turn my head to see whether I am proceeding in the right direction. I am amazed. There is nothing to be seen . . . I am alone; I can see nothing at all.'

Louis Blériot

FLYING INTO THE MIST

On Saturday 24 July 1909, Blériot and his rival Hubert Latham were camped on the French coastline near Calais. The weather had been poor, and Blériot had badly injured his foot in a previous flight. Nonetheless, he woke early and made a quick decision to take off at first light, jumping into his *Blériot XI* at 4.35 am and heading off into the mist.

It was a brave act, for Blériot was a poor navigator and was flying without a compass. Shortly after, rain and clouds had reduced his visibility to virtually nothing and he struggled to keep his aeroplane level. To his great relief, he saw the English coastline but then had to battle against gusting winds before dropping his plane to the ground, smashing his propellor and wheels in the process. It was an awkward landing, but Blériot had made the 39-km (24-mile) journey in just over 30 minutes, marking an extraordinary first in aviation.

Harriet Quimby

(1875–1912)

>>>>>>>>>>>>>>>>>>>>>>>>>

On 16 April 1912, Harriet Quimby – flying in an aeroplane borrowed from Louis Blériot – became the first female pilot to cross the English Channel. She was also America's first licensed pilot at a time when aviation was dominated by men, and women were unable to even vote, making her achievements all-the-more extraordinary.

A reporter for a New York magazine, Quimby took her first flying lessons in the spring of 1911. She had both natural talent and drive, and earned herself a pilot's license in just a few months.

She went on to fly with a demonstration team in front of large crowds and became known for her distinctive purple satin flying suit and hood. In March 1912, she decided to cross the English Channel and convinced Louis Blériot to loan her a monoplane.

UP AND AWAY

She took off from Dover, England, at 5.30 am on 16 April 1912 and was quickly surrounded by thick fog. She had never flown the plane before and the only instruments she had were a compass and a watch. Looking for an opening in the clouds, she descended, which caused the engine to backfire. Quimby began to fear she might have to ditch into the icy waters below. Thankfully, the engine recovered, and she was able to land on a French beach near to Hardelot.

Her heroic feat, however, was overshadowed by the sinking of the *Titanic* ship just two days before and Quimby received little fanfare. Less than three months later, she was tragically killed when she was thrown from her Blériot plane in front of five thousand spectators in Boston, USA.

> 'I'm going in for everything in aviation that men have done: altitude, speed, endurance, and the rest.'
>
> Harriet Quimby

Charles Lindbergh

(1902–1974)

Charles Lindbergh shot to fame when, in 1927, he flew nonstop over the Atlantic Ocean from New York to Paris in his *Spirit of St Louis* single-engine aeroplane. Lindbergh battled his way through thick fog and sleet, sometimes skimming the waves beneath, without a radio or even a clear view ahead to guide him on his daring feat of aviation.

THE CHALLENGE

Passionate about flying, trainee pilot Charles Lindbergh performed daredevil aerobatic stunts at fairs, and in 1924 signed up to the US Army Air Service. While working as an airmail pilot, Lindbergh decided to compete for a $25,000 prize (around $390,000 in today's money) to become the first person to fly solo nonstop from New York to Paris. In 1919, British pilots John Alcock and Arthur Brown had been the first to cross the Atlantic – flying from Canada to Ireland in an old First World War bomber – but no one had yet managed it alone. Others had lost their lives attempting the challenge, but Lindbergh was undeterred.

TAKE-OFF

With the financial support of businessmen from St Louis, USA, Lindbergh flew in a specially adapted monoplane, equipped with extra fuel tanks that obscured his view out of the cockpit. Taking off from Long Island on 20 May 1927, Lindbergh barely cleared the telegraph lines of the airfield before heading off towards the Atlantic.

SAFE RETURN

Reaching the vast ocean, darkness fell, storm clouds swirled all around and, to keep awake over the difficult flight, Lindbergh kept the side window open. After 33½ hours, he finally landed in Paris, where he was met by an ecstatic crowd. The flight proved that long-distance air travel was possible, and some four million people lined a parade route for Lindbergh on his return to New York.

IN THE HEADLINES AGAIN

Five years later, Lindbergh's name again hit the headlines when his infant son was kidnapped and murdered. On moving to Europe, he lost some public support when he advocated that the USA should stay out of the Second World War.

Amelia Earhart

(1897–1939)

›››››››››››››››››››››››››

Amelia Earhart had a passion for adventure and set many astonishing flying records, including becoming the first woman to fly on her own nonstop across the Atlantic Ocean. She also championed women in aviation and highlighted how difficult society made it for women to compete with men.

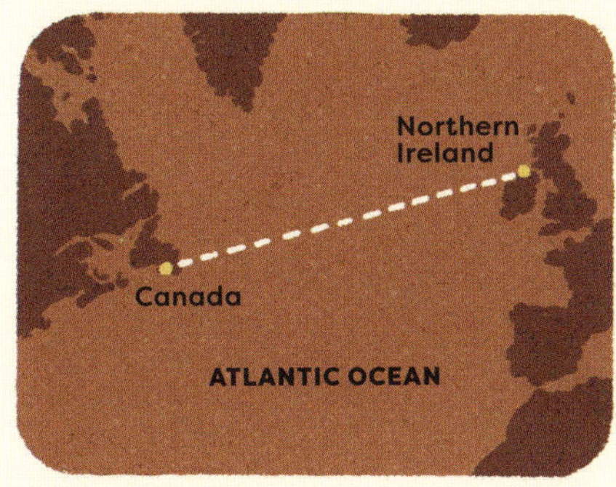

QUEEN OF THE AIR

In 1928 – just a year after Charles Lindbergh had flown from New York to Paris – Kansas-born Amelia Earhart flew as a passenger over the Atlantic. Earhart, however, was a trained pilot and she wanted to attempt the flight herself. In 1932, she completed the challenge, flying from Canada to Northern Ireland in a record time of 14 hours 56 minutes. Conditions were hazardous, with thick sea fog and ice forming on her plane's wings, and she was forced to land in a cow field.

ROLE MODEL

At a time when few women had the opportunity to fly, Earhart's transatlantic flight marked an incredible achievement and brought her worldwide fame. Earhart went on to become the first woman to fly across the United States. She also formed an organisation for female pilots and encouraged women to pursue challenges and opportunities in life.

'I'm quite aware of the hazards . . . I want to do it because I want to do it. Women must try to do things as men have tried. When they fail, their failure must be but a challenge to others.'

Amelia Earhart

AMELIA VANISHES

In 1937, Amelia set out to become the first woman to fly around the world. Tragically, during the last leg of the journey, her plane disappeared over the Pacific Ocean. It's thought she ran out of fuel, although the wreckage of her plane has never been found.

The Space Race

Having explored so much of the planet, people now looked to venture beyond Earth and into outer space. Humans had marvelled at the night sky for thousands of years, but few could have believed that people would one day travel there. In the 1950s, the USA and the former Soviet Union began to compete against each other to make this dream a reality.

Surveyor 1 probe approaching the Moon

ROCKET MEN

The first rockets were used as weapons in China as early as the 12th century. During the 20th century, scientists began to develop rockets that had enough force to escape the pull of Earth's gravity and to blast into space, but they were not used for space exploration until after the Second World War. The German V-2 rocket could travel at supersonic speed over long distances and, in 1944, a V-2 reached an altitude of 175 km (109 miles), making it the first man-made object to reach space.

ANIMAL ASTRONAUTS

After the war, scientists continued to explore the possibilities of space travel. In 1947, the USA launched a V-2 rocket into space carrying fruit flies to see how living creatures reacted to the speed, altitude and cosmic radiation. The following year, an American V-2 carried a Rhesus monkey called Albert into space. In 1957, the Soviet Union launched the first satellite, Sputnik 1, followed by a second satellite a month later carrying the dog Laika. Found wandering the streets of Moscow, Laika went on to become the first living creature to orbit Earth. Laika did not make it back alive, but she proved that animals – and potentially humans – could travel in space.

Laika, the first dog in space

COLD WAR

After the Second World War, tension developed between the two superpowers, the Soviet Union and the USA. This became known as the Cold War, so-called because neither country directly attacked each other. The Space Race formed part of the Cold War. To prove how strong they were, each country built up their weapons and developed technology, competing to be the first to send a spacecraft and person to the Moon.

SPACE MONKEYS

In 1958, the USA launched their own satellite, and in the following year sent two monkeys into space on a rocket. Each monkey was fitted into a special space suit and helmet, contained in a capsule. Their 16-minute journey involved nine minutes of weightlessness, before they dropped back down into the Atlantic Ocean. One of the monkeys, Able, died, but the other, Miss Baker, survived and became the first animal to safely return from space. The two animals became famous, and Miss Baker received hundreds of letters from school children every day.

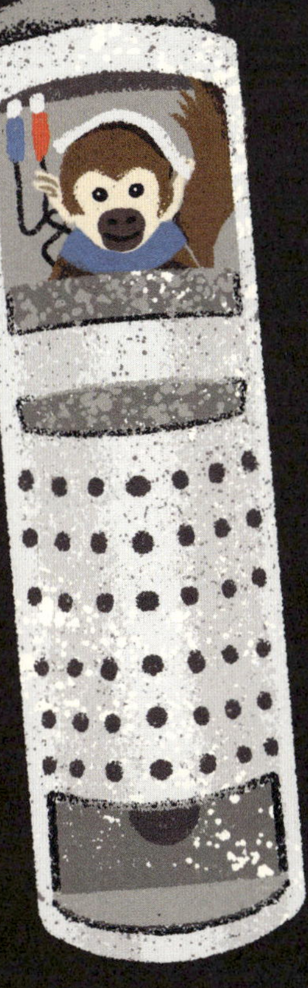

Miss Baker – the first animal to safely return from space

MOON STRUCK

The Soviet Union succeeded in putting the first man, Yuri Gagarin, into space in 1961 (p114), followed by the first woman, Valentina Tereshkova, in 1963 (p115). In 1966, both the Soviet Union and the USA sent unmanned spaceships to the Moon, which landed successfully. The US probe Surveyor 1 took photographs of the Moon's surface, which provided vital information for the eventual manned Moon landing. The race was now on to get the first person on the Moon.

A Vostok rocket blasting into space

> 'We choose to go to the Moon in this decade and do the other things, not because they are easy, but because they are hard.'
>
> US president John F. Kennedy

Yuri Gagarin

(1934–1968)

>>>>>>>>>>>>>>>>>>>>>>>>

The Soviet Union claimed the first victory in the Space Race when, in 1957, they put the spacecraft Sputnik 1 into orbit. Four years later, it sent the first human, Yuri Gagarin, into space. Travelling in a tiny capsule attached to a powerful rocket, Gagarin blasted into space and flew around the Earth. The journey marked one of the most significant moments in space history and earned Gagarin worldwide fame.

LET'S GO!

At 9:07 am on 12 April 1961, 27-year-old Yuri Gagarin, a former fighter pilot with the Russian Air Force, rocketed into space. Setting off from Baikonur Cosmodrome, now in Kazakhstan, Gagarin travelled in a capsule-like spacecraft called Vostok 1. Although Gagarin was fully trained, no one knew if the mission would be a success or failure.

I see Earth! It is so beautiful.

Yuri Gagarin

Just after lift-off, Gagarin famously called out, '*Poyekhali!*' ('Let's go!'). On entering space, he completed one orbit of Earth at speeds of about 27,400 km (17,000 miles) per hour – ten times faster than a rifle bullet. After 1 hour 48 minutes, he re-entered the Earth's atmosphere and, as planned, ejected from the spacecraft and parachuted to the ground.

HISTORIC JOURNEY

A local farmer and her daughter saw Gagarin floating down in his parachute and were shocked by the sight of the strange figure dressed in an orange space suit and large white helmet. He told them not to be afraid and asked if they could direct him to the nearest telephone. As soon as the world heard of his historic journey, Gagarin was hailed as a hero.

Valentina Tereshkova

(1937–)

›››››››››››››››››››››››››

The Soviet Union also wanted to send the first woman into space. This was achieved in 1963 when Valentina Tereshkova lifted off without fault in the Soviet spacecraft Vostok 6 and orbited Earth 48 times. Her record-breaking journey occurred a staggering 20 years before the USA sent their first female astronaut into space.

SOVIET SPACE PROGRAMME

Inspired by the space flight of Yuri Gagarin, Valentina Tereshkova applied to be a cosmonaut while working as a textile worker in a local factory. Although she had no pilot training, she was a passionate parachutist, a rare skill for women in the 1960s. She was accepted for intensive training at the age of 24 and completed her mission two years later, making her also the youngest person ever to fly into space.

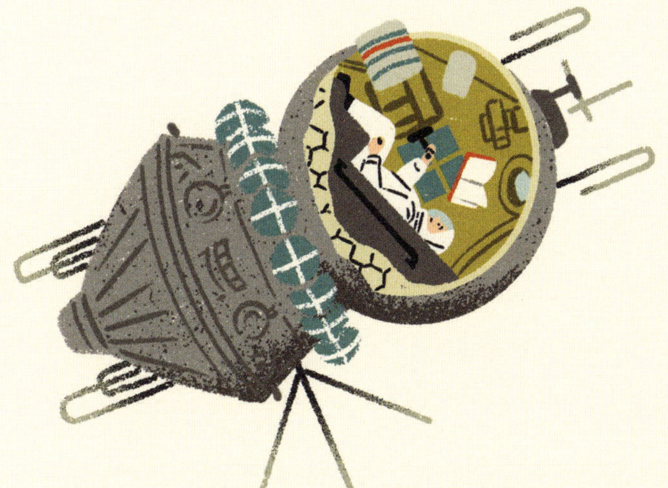

SEAGULL MISSION

On 16 June 1963, Tereshkova was launched into orbit aboard the capsule spacecraft, Vostok 6. Her flight lasted 2 days, 23 hours and 12 minutes and during that time she completed 48 orbits of the Earth. Unlike Gagarin, Tereshkova had manual controls on board, and she also communicated with another spacecraft, Vostok 5, using the radio call sign *Chaika* ('seagull').

Tereshkova ejected from Vostok 6 and parachuted down to Earth, landing on 19 June. After her historic flight, she received multiple awards across the world and went on to have an influential political career both in the Soviet Union and abroad.

> 'A bird cannot fly with one wing only. Human space flight cannot develop any further without the active participation of women.'
>
> Valentina Tereshkova

Neil Armstrong

(1930–2012)

On 20 July 1969, Neil Armstrong stepped off the lunar landing module Eagle and became the first human being to walk on the Moon. Broadcast live on television, half a billion people gazed in wonder at what was the pinnacle of human exploration. Buzz Aldrin accompanied Armstrong, and subsequent US Apollo missions saw another ten astronauts walk on the Moon.

TEENAGE PILOT

Born in Ohio, USA, Neil Armstrong developed a passion for flight at an early age. Gaining a student's pilot licence at the age of 16, he went on to study aeronautical engineering before flying fighter jets for the US Navy. He later worked as a test pilot, flying high-speed aircraft such as the rocket-powered X-15, which could reach speeds of 7,200 km (4,400 miles) per hour.

EAGLE HAS LANDED

In 1966, Armstrong led the flight of the US spacecraft Gemini 8 into space. He was then selected as commander for a planned Apollo 11 Moon landing mission, alongside astronauts Edwin E. (Buzz) Aldrin Jr and Michael Collins. On 16 July 1969, a rocket blasted the Apollo 11 spacecraft into space. As it neared the Moon, Armstrong and Aldrin moved into one of Apollo's three parts, the landing module Eagle. Collins remained in the Columbia command module. Armstrong guided Eagle down on to the Moon's surface with only 25 seconds of fuel left.

SMALL STEP

Shown live on television across the world, Armstrong descended the stairs of Eagle and took his first steps on the Moon's surface. Nineteen minutes later, Aldrin joined him and, for around two hours, they took photographs, collected soil samples and marvelled at the amazing sight of Earth (which Armstrong said looked beautiful). They also planted the US flag on the surface. They then returned to Eagle, lifted safely off the Moon, joined Columbia and returned to Earth, splashing into the Pacific Ocean after more than eight days in space.

> 'That's one small step for [a] man, one giant leap for mankind.'
>
> Neil Armstrong as he put his foot onto the Moon

APOLLO ASTRONAUTS

Between 1969 and 1972, another ten astronauts went on to land on the Moon, all of them on US Apollo missions. Apollo 12 made a successful Moon landing in November 1969, but, five months later, disaster struck the Apollo 13 mission when an oxygen tank exploded. The crew never made it to the Moon but managed to return safely to Earth. In 1971, Alan Shepard (who in 1961 had been the second person and first American to travel into space) commanded the Apollo 14 mission. Television images famously showed him on the Moon's surface hitting golf balls with a club. The Apollo 15 mission took a roving vehicle to explore the Moon's surface, and on the final Moon landing, Apollo 17, astronauts spent four days on the Moon. The Apollo missions were regarded as one of the greatest technological achievements in history.

To date, 12 astronauts have landed on the Moon, all between 1969 and 1972:

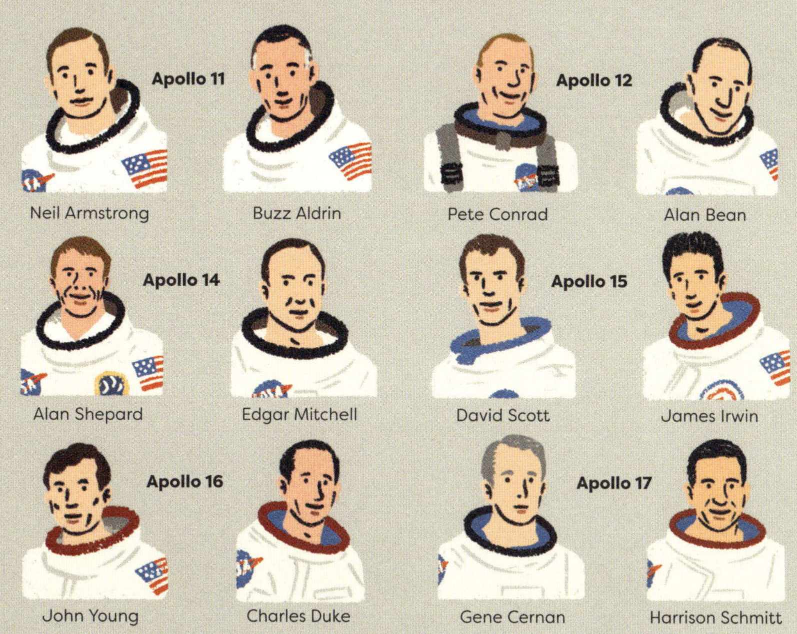

VICTORY

On their return, Armstrong, Aldrin and Collins were considered great heroes across the world. The mission had also secured America's victory in the Space Race.

Space Pioneers

The last 60 years of space exploration have seen spectacular firsts as astronauts pushed the boundaries of space travel, paving the way for future generations. Some were brave enough to leave their spacecraft and float freely in the vast expanse of space. Women too now form a critical part of space missions, and astronauts from across the world have ventured into the cosmos.

WALKING IN SPACE

The first person to leave a spacecraft and float in space, or spacewalk, was Soviet cosmonaut Alexei Leonov (1934–2019). On 18 March 1965, on the Voskhod 2 mission, Leonov exited the air lock and stepped out into space, attached to his spacecraft only by a tether.

His milestone nearly turned to tragedy, however, when his space suit ballooned during his spacewalk, meaning he couldn't get back through the airlock into his spacecraft. He managed to make it inside the craft, but only after letting out some oxygen from his suit, causing him to suffer from decompression and to sweat so much that it sloshed around his suit. The spacecraft then went into a spin. Leonov and his fellow cosmonaut Pavel Belyayev eventually managed to stabilise the craft and land back on Earth, albeit wildly off course in a thick forest.

Nineteen years later, on 7 February 1984, US astronauts Bruce McCandless II and Robert Stewart left the safety of the Challenger space shuttle and embarked on the first untethered spacewalk. They travelled more than 90 m (300 ft), entirely unconnected to the shuttle, using a jetpack device that allows astronauts to manoeuvre themselves in space.

'It was so quiet I could even hear my heartbeat. I was surrounded by stars and was floating without much control.'

Alexei Leonov

CUBAN PIONEER

The first person of African heritage to go into space was Cuban pilot Arnaldo Tamayo Méndez (born 1942). He was launched into space on board the Soyuz 38 spacecraft on 18 September 1980 with Soviet cosmonaut Yuri Romanenko. During his 8-day mission he docked with the Salyut 6 space station, conducted various experiments and orbited Earth 124 times in 7 days and 20 hours.

FIRST AFRICAN AMERICAN

The third flight of the US space shuttle Challenger on 30 August 1983 saw the first African American, Guion 'Guy' Bluford (born 1942), blasted into space. The former fighter pilot's mission included the deployment of an Indian communications satellite. He would travel on Challenger for a second time in 1985 and later in 1991 and 1992 on board the US space shuttle Discovery.

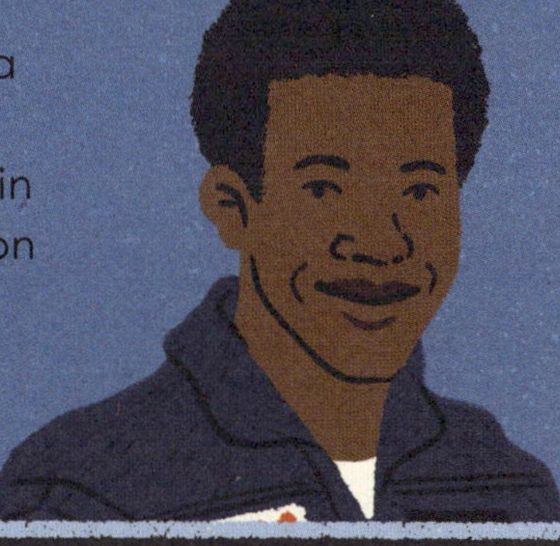

FIRST AMERICAN WOMAN

1983 would see the first American woman in space when physicist Sally Ride (born 1951) joined the crewmembers of the Challenger space shuttle on 18 June 1983. She was selected for the space mission in 1978, the first year the USA accepted women for astronaut training.

FIRST BLACK WOMAN

The flights of Sally Ride and Guion Bluford had inspired engineering graduate and medic Mae Jemison (born 1956) to apply to the US space programme. In 1992, she served as a mission specialist aboard the space shuttle Endeavour, becoming the first black woman to travel into space.

'I've come to appreciate the planet we live on. It's a small ball in a large universe. It's a very fragile ball but also very beautiful. You don't recognise that until you see it from a little farther off.'

Guion Bluford

Future Exploration

Humans continue to explore uncharted worlds both on our planet and beyond. Robotic spacecraft and telescopes beam back images of alien worlds – and ordinary people, not just astronauts, will experience the wonders of space travel. On Earth, deep-sea exploration is helping us better understand the evolution of life and how we can protect our fragile planet for future generations.

THE SOLAR SYSTEM

In recent years, scientists have made great leaps in learning about Earth's Solar System. Spacecraft have orbited the planets Mars and Mercury, and flown near to Pluto, and a robotic module landed on a comet that was moving 40 times faster than a speeding bullet. February 2021 saw three different missions arrive on Mars. First, the United Arab Emirates's Hope orbiter entered orbit. China sent an orbiter and a robotic vehicle, known as a rover, to land there, the first in a series of planned missions by China to explore the Solar System. The USA also landed a rover to look for signs of alien life and tested a helicopter, which proved itself a useful tool for future planet exploration.

The Mars Perseverance rover and the helicopter Ingenuity searching for signs of past life

FAR SIDE OF THE MOON

In 2019, China landed a robotic spacecraft on the 'far side' of the Moon, a previously unexplored part of the Moon that is never visible from Earth. The USA also plan to land the first woman and the next man on the Moon by the mid-2020s.

THE BEGINNINGS OF THE UNIVERSE

The James Webb Space Telescope was launched into space on Christmas Day 2021. The biggest and most powerful telescope ever built, it will travel 1.5 million km (1 million miles) away from Earth. It will transform our knowledge of the Universe, allowing scientists to look further into space than ever before, search for alien life and see light from the earliest stars, and the Universe when it was young.

FROM DEEP SPACE TO DEEP SEA

Just as much of the Universe is unknown to humans, there are vast areas of the Earth's seabed that we know nothing about. Undersea drones have the potential to transform our understanding of the ocean floor and an international scientific team aims to map the entire floor of the Earth's oceans by 2030. The discovery that strange animals could live in the depths of the sea has emboldened scientists to open their minds to the possibility that similar life forms could exist without sunlight on other planets. Ocean exploration could also help scientists better understand how the changes in weather and climate affect our planet.

SPACE TOURISTS

Most recently, private citizens and businesses have funded space missions. Jeff Bezos, owner of Blue Origin and founder of Amazon, and Sir Richard Branson, CEO of Virgin Galactic, both flew to the edge of space in 2021. These companies expect to begin regular space missions, offering tourists a few minutes of weightlessness before returning to Earth. One of Blue Origin's more famous passengers was the actor William Shatner, who played a fictional space explorer in *Star Trek*. Blasting off from a desert in Texas, Shatner, at the age of 90, became the oldest person to travel to space.

'I was once a child with a dream looking up to the stars. Now I'm an adult in a spaceship looking down to our beautiful Earth. To the next generation of dreamers: if we can do this, imagine what you can do.'

Richard Branson

AN INTERVIEW WITH

A Modern-day Explorer

EMILY FORD

In 2021, the then 28-year-old Emily Ford became the first woman, the first person of colour and only the second person ever to complete the Ice Age Trail in winter. This involves an incredibly arduous trek over 1,900 km (1,200 miles) of icy terrain in Wisconsin, USA. The trail traces the edge of glaciers from the last Ice Age. With just an Alaskan husky called Diggins for company, Emily completed her gruelling adventure in 69 days.

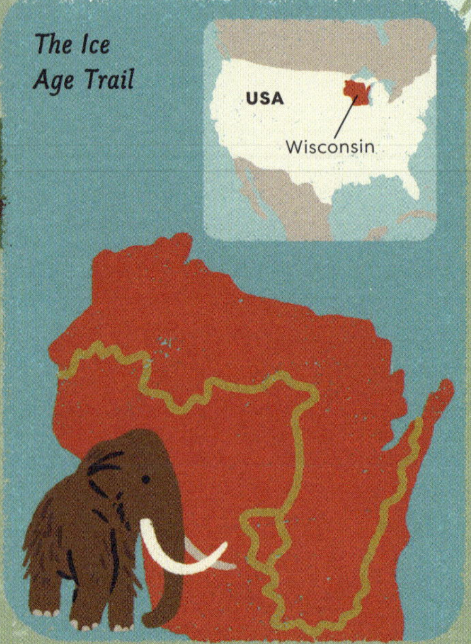

WHAT INSPIRED YOU?

At first, I just wanted to hike a really long distance. Then I realised I also wanted to show others that anyone can explore the great outdoors – no matter what you look like or what your abilities are.

WHAT PREPARATIONS DID YOU MAKE?

First, I made a list of all the things I would need: food, clothing and cooking and camping gear. I mapped each day on the route and planned my resupply spots so that I didn't have to carry everything with me. The Ice Age Trail involves 800 km (500 miles) of road walking, so I practised doing long walks on the road with my dog.

WHAT WERE THE HIGHLIGHTS?

One of the highlights of the trip for me was sleeping outside and waking up surrounded by beautiful forests covered in frost. Meeting people on the trail was another highlight. I was often greeted with food or hot drinks, and I encountered a lot of kindness. When I finally completed the trek, I was welcomed by a large crowd, cheering me on. Despite my exhaustion, it was an exhilarating feeling and I was overwhelmed that so many people had turned up.

WHAT ADVICE WOULD YOU GIVE TO A YOUNG EXPLORER?

Today, most places have already been explored by somebody, but I'd argue that every inch of every place has yet to be found. The Earth is ever changing and shifting. When we re-explore the Arctic and Antarctic regions, it is completely different to what our predecessors saw. After all, our Earth is warmer than theirs!

Anyone can become an explorer. You just need to find an adventure that interests you, then do your research and ask lots of questions. If you experience a setback or challenge, there's usually something you can learn from it, and you will get back towards your goal in time. Just remember, while following your dreams, always have empathy for others and be mindful of the places you visit and the people who live there.

INSPIRATION

One explorer who inspired me is Matthew Henson – who may have been the first person to make it to the North Pole (p90). He did everything he could to overcome the barriers he faced as a black man. He was smart and knew how to work harder than anyone else around him so he could keep going on expeditions. Another inspiration for me is the polar explorer Ann Bancroft (p99). Reading her stories reminds me to keep pushing the boundaries for women and minorities. When she completed her expeditions, many companies would have rather sponsored a man than a woman. This is still true for some today, but because of women like Ann, it is much easier.

GLOSSARY

ABORIGINAL AUSTRALIANS
People who have been living in Australia since well before European settlers arrived.

AIRSHIP
A large, engine-powered aircraft with a body filled with lighter-than-air gas.

ALTITUDE
The height of something above the level of the sea or ground.

AMERICAN WEST
The westernmost states of the USA.

ANCESTOR
A member of your family who lived long ago.

ANTARCTICA
The most southernmost continent and landmass on Earth, which contains the South Pole.

AQUALUNG
A breathing system that feeds air to a diver from an oxygen tank, allowing them to stay underwater.

ARCHAEOLOGY
The study of the past by digging up sites and analysing bones, tools and physical remains.

ARCTIC
Region located at the northernmost part of Earth. It includes the Arctic Ocean, the North Pole, other landmasses and adjacent seas.

ASTRONOMER
A scientist who studies stars, planets and other objects in outer space.

AVIATION
The flying of aircraft in the skies.

AZTECS
One of the peoples who lived in Mexico before the Spanish colonists arrived.

BATHYSCAPE
A special vessel used for deep-sea dives.

BEDOUIN
An Arab person who lives in the desert and moves from place to place.

BOTANIST
A person or scientist who studies plants.

BYZANTINE
Relating to the empire that emerged out of the eastern half of the Roman Empire.

CARAVAN
A group of people travelling across a desert, often with camels or animals.

CARTOGRAPHER
A person who draws or produces maps.

CIRCUMNAVIGATE
To travel all the way around the world.

COLD WAR
A period of tension and hostility between the superpowers of the Soviet Union and the United States that lasted from the Second World War to the early 1990s.

COLONY
A settlement in a country or land built by people (called colonists) from another country or land.

COMET
A lump of rock, gas or dust that travels around the Sun.

CONQUISTADORS
The Spanish explorers and soldiers who invaded and conquered much of the Americas.

COMPASS
An instrument that points out directions, such as north, south, east and west.

COSMONAUT
A Russian or Soviet astronaut.

COSMOS
The Universe.

CURRENT
A body of air or water moving in one direction.

DHOW
A sailing boat commonly used by Arab sailors in the Indian Ocean.

DISCRIMINATION
To treat someone unfairly for a particular reason, such as their race, gender or religion.

DISPOSSESSION
The act of taking someone's land or property away from them.

DRONE
An unmanned vehicle that can fly or explore the ocean depths.

ECOSYSTEM
Any area where living things, such as plants and animals, interact with non-living things, like soil and air.

EQUATOR
An imaginary line running around the middle of Earth.

EVOLUTION
The process by which living things gradually change over millions of years.

FROSTBITE
Damage to skin and the body when exposed to freezing temperatures.

GLIDER
A type of aircraft that doesn't have motors.

GEOLOGIST
A scientist who studies rocks and the structure of the Earth.

HAJJ
A Muslim pilgrimage (journey) to the sacred city of Mecca in Saudi Arabia.

HOT-AIR BALLOON
A simple aircraft made up of a large balloon filled with hot air and a basket underneath.

HYDROTHERMAL (OR DEEP-SEA) VENTS
Openings in the sea floor which spray out hot, mineral-rich water.

INCA
A people who controlled a vast empire along the west coast of South America before European colonisers arrived.

INDIGENOUS PEOPLES
A group of people who lived in a region before settlers or colonists arrived.

INUIT
Native peoples of northern Canada, Greenland, Alaska and the Arctic.

LATITUDE
Imaginary horizontal lines running at equal distance above and below the Equator.

LOCOMOTIVE
An engine that pushes or pulls a train on rails.

LONGITUDE
Imaginary lines running from the North Pole to the South Pole on Earth.

LONGSHIP
A long, narrow ship sailed by Vikings in rough oceans and shallow waters.

MALARIA
A disease spread from person to person by mosquitoes.

MARITIME
Related to shipping, the seas and oceans.

MIGRATION
The movement of people from one place to another, often in large numbers.

MING DYNASTY
A family of emperors who ruled China from 1368 to 1644.

MISSIONARY
A person who is sent to a place to spread a religious faith.

MONOPLANE
An aeroplane with one pair of wings.

MOUNTAIN MEN
Men, also known as fur trappers, who travelled across the American wilderness, hunting, laying traps and trading fur and skins of animals.

MUTINY
A refusal by a group of soldiers or sailors to obey the orders of a leader or captain.

MYTH
Relating to the old, fictional stories about a place or culture.

NATIVE
A person who comes from, or is born in, a particular place.

NATURALIST
A scientist or person who studies nature, especially plants and animals.

NAVIGATE
To find a way from one place to another.

NATIVE AMERICAN
The first groups of people to live in North and South America.

NORTH POLE
The most northerly point on Earth, which sits on constantly shifting ice in the middle of the Arctic Ocean.

NORTHWEST PASSAGE
A sea route north of Canada connecting the Atlantic and Pacific Oceans.

ORBIT
To go around a star, planet or a moon in space.

OTTOMAN
Relating to the Ottoman Empire – its capital was Constantinople (modern-day Istanbul in Turkey).

PENNY-FARTHING
An early bicycle, noticeable for its large front wheel.

PHARAOH
A ruler in ancient Egypt.

PILGRIMAGE
A long journey to a holy site.

ROVER
A robotic vehicle that explores the surface of a planet, comet or moon.

SAGA
An old Icelandic story of adventures and heroic deeds.

SAKOKU
A period of isolation for Japan, when no foreigner or Japanese person could leave or enter the country.

SAMURAI
A highly skilled Japanese warrior.

SARCOPHAGUS
A stone container for a coffin or body.

SATELLITE
A small object or spacecraft that goes around a larger object or planet in space.

SCURVY
A disease often suffered by sailors or polar explorers who didn't eat enough vitamin C, which is found in citrus fruit and vegetables.

SHERPAS
People who live in the Himalayan mountains, known for their mountaineering skills.

SLAVE TRADE
The selling of people as slaves, which was at its height in the 18th century when Europeans enslaved Africans and transported them to the Americas.

SMALLPOX
An infectious disease that European explorers brought to the Americas, killing millions.

SNOWMOBILE
A small vehicle used for travelling on snow.

SONGLINES
Also known to Australian Aboriginal peoples as dreaming tracks, these describe the features and routes through the landscape.

SOUTH POLE
The most southerly point on the Earth, found in Antarctica, the last continent to be discovered by human beings.

SPACE RACE
Competition between the Soviet Union and the USA to dominate space exploration in the 1960s and 1970s.

SPECIES
A group of animals, plants or other living things that share common characteristics.

SPICE ROUTES
A network of maritime trade routes connecting Europe and Asia so that people could buy and transport spices. The land routes were also called the Silk Road.

SUBMERSIBLE
A vehicle that can operate underwater.

TELEGRAPH
An old-fashioned system of sending messages over long distances using wires.

TREASURE SHIP
A ship loaded with gold, silver and other precious cargo. It was also a type of large wooden ship in the fleet of the Chinese admiral Zheng He.

WOOLLY MAMMOTH
An extinct elephant-like animal that lived in Europe until around 4,000 years ago.

INDEX

A

Aboriginal Australians 6, 36–37, 68, 69, 70–71
Age of Discovery 20–21
airships 106, 107
Al-Idrīsī, Muhammad 22
Alcock, John 110
Aldrin, Edwin 'Buzz' 116
Amundsen, Roald 89, 91, 92, 94–95
Anarson, Ingólfr 15
ancient civilizations 78–79
    Aztecs 54
    Egyptians 10, 78
    Incas 54, 55, 58, 78, 79
    Phoenicians 10, 11
    Polynesians 13
Apollo Moon missions 116–117
aqualungs 43, 44
Armstrong, Neil 116–117
Arnesen, Liv 99
astrolabes 23
automobiles 85
aviation 106–107

B

Baker
    Florence 75
    Samuel 75
Balmat, Jacques 100
Bancroft, Ann 99, 123
Banks, Joseph 34
Bar Sauma, Rabban 51
Baret, Jeanne 32–33
Barth, Heinrich 72
Barton, Otis 43
bathyscapes 45
bathyspheres 43
Beckwourth, James 62–63
Beebe, William 43
Bell, Gertrude 76–77
Belyayev, Pavel 118
Bennett, Floyd 91
Bingham, Hiram 79
Bird, Isabella 66–67
biremes 11
Bisland, Elizabeth 83
Blériot, Louis 108, 109
Bluford, Guion 'Guy' 119
Bly, Nellie 82–83
Borchgrevink, Carsten 92
Bowers, Henry 95
Bransfield, Edward 92
Brown, Arthur 110
Bungaree 36–37
Burckhardt, Johann 78
Burke, Robert O'Hara 68–69
Burton, Charles 98
Byrd, Richard 91

C

Cabot, John 21, 88
Cameron, James 43
canoes
    double 12
    outrigger 12
Carter, Howard 79
Cartier, Jacques 21, 88
Cayley, George 107
Charbonneau, Toussaint 61
Charles, Jacques 106
Chichester, Francis 40
Chojnowska-Liskiewicz, Krystyna 40
chronometers 23
Clark, William 60–61
Cochran, Elizabeth Jane. See Bly, Nellie
Collins, Michael 116
Columbus, Christopher 16, 20, 24–25, 51, 54, 88
Commerson, Philibert 32–33
compasses 19, 23
conquistadors 54–55
Cook
    James 21, 34–35, 36, 92
    Thomas 65
Cortés, Hernán 54
Cousteau, Jacques 43, 44

D

Da Gama, Vasco 20, 26, 27
Darwin, Charles 38–39
De Bougainville, Louis-Antoine 32, 33
De León, Juan Ponce 55
De Narváez, Pánfilo 56, 57
De Orellana, Francisco 58–59
De Saussure, Horace-Bénédict 100
deep-sea diving 42–43
Dekker, Laura 41
dhows 27
Dias, Bartolomeu 20
diving bells 42
diving suits 42
Drebbel, Cornelis 42

E

Earhart, Amelia 111
Egingwah 90
Elcano, Juan Sebastián 28, 29
Erik the Red 15, 16
Erikson, Leif 15, 16–17
Estevanico 56–57
Evans, Edgar 95

F

Fiennes, Ranulph 98
Flinders, Matthew 36–37
Ford, Emily 122–123
Franklin, John 88, 89
Frobisher, Martin 88

G

Gabart, Francois 41
Gagarin, Yuri 113, 114, 115
Gagnan, Émile 43, 44
Giffard, Henri 107
gliders 106, 107
Gray, Charles 68–69

H

Hall, Idris Galcia. See Wanderwell, Aloha
Halley, Edmund 42
Harrison, John 23
He, Zheng 18–19
Heezen, Bruce 46
Henson, Matthew 90-91, 123
Hillary, Edmund 100, 102–103
hot-air balloons 45, 106
Hudson, Henry 88

I

Ibn Battuta 52–53
Ibn Mājid, Ahmad 27
Ice Age Trail 122–123
Inuit 89, 90, 95
Irvine, Andrew 102

J

James Webb Space Telescope 121
Jemison, Mae 119
jetpacks 118
Joyon, Francis 41

K

Kaltenbrunner, Gerlinde 101
'kelp highway' 7
King, John 68–69
Kingsley, Mary 74
kites 106, 107
Knox-Johnston, Robin 40
Kopchovsky, Annie. See Londonderry, Annie
Kwasi, Graman 39

L

Latham, Hubert 108
Lawrence, T. E. (Lawrence of Arabia) 77
Leonov, Alexei 118
Lethbridge, John 42
Lewis, Meriwether 60–61
Lilienthal, Otto 107
Lindbergh, Charles 110, 111
Livingstone, David 73
Londonderry, Annie 84–85
longships 14

M

MacArthur, Ellen 41
MacDonald, Sheila 100
Magellan, Ferdinand 22, 28–29
Mallory, George Leigh 102
mapmakers 22–23
McCandless II, Bruce 118
McClure, Robert 89
Mendez, Arnaldo Tamayo 119
Mercator, Gerardus 23
Messner, Reinhold 101
Meyer, Hans 100
Mongol Empire 50, 51
Montgolfier
 Etienne 106
 Joseph 106
monoplanes 108, 109, 110
Moon landings 113, 116–117
Mouhot, Henri 78

N

Native Americans 7, 25, 56, 60–61, 63, 65
navigation 11, 13, 15, 19, 23, 27, 70–71
Norgay, Tenzing 100, 102–103

O

Oates, Lawrence 95
Ooqueah 90
Ootah 90, 91
Ortellius, Abraham 23
Outen, Sarah 85

P

Paccard, Michel-Gabriel 100
Palmer, Nathaniel 92
Peary, Robert 90–91
Perham, Michael 41
Piccard
 August 45
 Jacques 43, 45
Pizarro
 Francisco 55, 58–59
 Gonzalo 58
Plain, Steven 101
Polo, Marco 50–51
Ptolemy 22

Q

Quimby, Harriet 109

R

railways 64–65
Raven-Flóki. See Vilgerdarson, Flóki
Ride, Sally 119
Romanenko, Uri 119

S

Sacagawea 60–61
Scott, Robert Falcon 93, 94–95
Seegloo 90
Shackleton, Ernest 93, 96–97, 99
Shepard
 Alan 117
 Oliver 98
sherpas 102
Silk Road 10, 50
Slocum, Joshua 40
Space Race 112–113, 114, 117
space tourism 121
Spice Routes 10
Stanley, Henry 73
Stark, Freya 79
Stevens, Thomas 84
Stewart, Robert 118
submarines 42
submersibles 42, 43
 Alvin 43, 47
 Deepsea Challenger 43
Sunderland, Zac 41

T

Tabei, Junko 100, 101
Tereshkova, Valentina 113, 115
Tharp, Marie 43, 46
Thorbjarnardóttir, Gudrid 17
treasure ships 18
Tsunenaga, Hasekura Rokuemon 30–31
Tupaia 35

V

Van Dover, Cindy Lee 47
Vikings 14–15
Vilgerdarson, Flóki 15
Villas-Bôas
 Cláudio 80–81
 Leonardo 80–81
 Orlando 80–81
Von Bellingshausen, Fabian Gottlieb 37, 92
Von Humboldt, Alexander 39
Von Zeppelin, Ferdinand 107

W

Waldseemüller, Martin 22
Walsh, Don 45
Wanderwell, Aloha 85
Wills, William John 68–69
Wilson, Edward 93
Wright
 Orville 107
 Wilbur 107

Z

zeppelins 107